*The*
*Pleasures of*
# Pi,e
*and Other Interesting Numbers*

# The
# *Pleasures* of
# Pi,e
## and Other Interesting Numbers

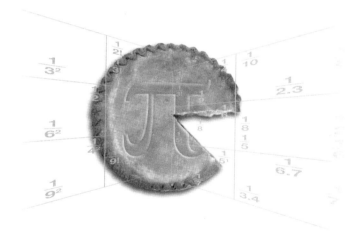

# Y E O Adrian

M.A., Ph.D. (Cambridge University)
Honorary Fellow, Christ's College, Cambridge University

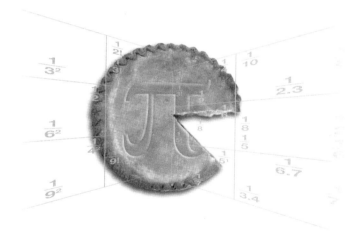 **World Scientific**

NEW JERSEY · LONDON · SINGAPORE · BEIJING · SHANGHAI · HONG KONG · TAIPEI · CHENNAI

*Published by*

World Scientific Publishing Co. Pte. Ltd.

5 Toh Tuck Link, Singapore 596224

*USA office:* 27 Warren Street, Suite 401-402, Hackensack, NJ 07601

*UK office:* 57 Shelton Street, Covent Garden, London WC2H 9HE

**British Library Cataloguing-in-Publication Data**
A catalogue record for this book is available from the British Library.

First published 2006
Reprinted 2007

ISBN 981-270-078-1
ISBN 981-270-079-X (pbk)

Printed in Singapore by B & JO Enterprise

*Dedicated To*

*Kathryn and Rebecca*

**Rebecca's "passwords" for access to computer games**

$$\pi = \frac{circumference}{Diameter}$$

$$\pi = 3.1415926535$$

$$\frac{\pi}{4} = \frac{1}{1} - \frac{1}{3} + \frac{1}{5} - \frac{1}{7} + \frac{1}{9} \cdots$$

$$\frac{\pi^2}{6} = \frac{1}{1^2} + \frac{1}{2^2} + \frac{1}{3^2} + \frac{1}{4^2} + \frac{1}{5^2} \cdots$$

02.04.06

Rebecca
Age : 6

**Kathryn's "password" for access to computer games**

$$\underline{e}$$

$$e = 1 + \frac{1}{1!} + \frac{1}{2!} + \frac{1}{3!} + \frac{1}{4!} + \cdots$$

20.4.06

Kathryn

Age: 4

A
was once an apple pi,e
Pidy Widy
Tidy Pidy
Nice Insidy
Apple pi,e

**Edward Lear** (1818–1888)

&

A thing of beauty is a joy forever
Its loveliness increases
It will never pass into nothingness

**John Keats** (1795–1821)

&

The most beautiful thing you can experience
is the mysterious
It is the source of all true art and science
He ... who can no longer pause to wonder
and stand rapt in awe
is as good as dead

**Albert Einstein** (1879–1955)

&

Mathematics is like checkers
in being suitable for the young
not too difficult, amusing
and without peril to the state

**Plato** (~429–347 BC)

❦

Equations are more important to me
because politics is for the present
but an equation is for eternity

**Albert Einstein** (1879–1955)

❦

In nature's infinite book of secrecy
A little I can read

**William Shakespeare** (1564–1616)

❦

# Foreword

For many, both young and old, mathematics strikes cold fear. Yet, mathematicians make no apology for this nor do they attempt to bring things to the level of mere mortals. As one who has spent many years in education, I am struck by how effortlessly Dr Y E O Adrian's wonderful book bridges this divide, making mathematics fun and accessible even to the math averse.

Written in a warm and graceful style, this unusual book has the feel of a journey through time interspersed with numerous historical references and interesting anecdotes. Using infinite series as a thread, the book reveals the beauty and elegance as well as the intellectual challenges in mathematics. Building on the more familiar — arithmetic, algebra, and elementary functions — the book takes readers through less familiar terrain in mathematics. Along the way, it helps them experience the "magic" that emerges from the idea of infinity.

The author's exposition exemplifies an experiential approach, navigating through arithmetic and algebraic manipulations to reveal deeper patterns and beautiful symmetries. This is accomplished by building on existing knowledge in a cumulative manner to arrive at further results and insights in a refreshing way. In my own experience, I'd say that the five most remarkable symbols in mathemtics are $\pi$, $e$, $i$, $0$, and $1$, and they are beautifully treated in this book.

One unusual aspect of this book is the potential breadth of its readership. By focusing on a coherent set of ideas around the concept of infinity, *The Pleasures of pi,e* takes its readers through a tantalizing mathematical adventure. This book should appeal to the young from pre-teen to pre-university, in addition to grown-ups with an interest in mathematics.

Professor Shih Choon Fong
MS, PhD (Harvard)
President, National University of Singapore
Foreign Associate, US National Academy of Engineering
Foreign Honorary Member, American Academy of Arts and Sciences
Chevalier, Order of "Legion d'Honneur"

# Foreword

I have known Dr Y E O Adrian for a long time. His ability to develop a deep insight into a particular topic and to argue logically is an attribute few of us possess.

He has written an interesting book on infinite series, a subject which has attracted the attention of many mathematicians throughout the ages. The results of the summations of infinite series are often elusive and surprising. He has managed to point out very clearly the key features of each series, which is his major contribution.

The book has a good collection of infinite series related to $\pi$ and $e$, some well-known and others not commonly found. Anyone who has a serious interest in infinite series will find his book a good reference.

Another interesting feature is the numerous quotations by famous writers, philosophers, scientists and mathematicians from Plato to Albert Einstein. They provide a number of refreshing views of mathematics.

I find his book really fascinating.

Professor Cham Tao Soon
BE, BSc, PhD (Cambridge)
Fellow, Royal Academy of Engineering, UK
Member, Swedish Royal Academy of Engineering
Honorary Fellow, St. Catharine's College, Cambridge
University Distinguished Professor
(Former President, 1981–2002)
Nanyang Technological University, Singapore

# Preface

"This is mission impossible", said my friends, when they heard that I was writing a math book for the enjoyment of non-mathematicians and those who "hated math in school". They are right. Even Stephen Hawking and Roger Penrose, two of the most famous mathematicians in the 21st century, alluded in their books to the common belief that each mathematical equation in a book would halve its sales and readership. Still I persisted.

This book is written for the young and the young-at-heart. They respond more readily to beauty, patterns and symmetry, with awe and amazement. The young are always interested in what they see around them in the world. They wonder about how things can be the way they are. And they always ask the question "why?"

This book began life as a booklet, some twenty, thirty pages long, and was meant only for my granddaughters. Then it took on a life of its own, and grew and grew, as more and more equations were added.

The genesis of the book was prompted by my two granddaughters, Rebecca, 6 and Kathryn, 4. They love playing computer games. After short lessons with me, they would play the games again and again, until they master them. Some of the games require fairly complex heuristic algorithms (game plans), executed in logical systematic sequences. And yet the kids could master them fairly effortlessly. As I watched them playing enthusiastically, it occurred to me that young children might be able to master complex mathematical algorithms too, if they were similarly motivated.

Therefore I decided to teach Rebecca some mathematics. I mentioned $\pi$, explained what it is, and gave her the value of $\pi$ up to 10 decimal places. A few days later, I asked her what $\pi$ was. She gave me the answer, with its value correct for 9 digits out of 10, and in the right sequence. I then taught her the Liebniz-Gregory infinite series for $\frac{\pi}{4}$, and Euler's series for $\frac{\pi^2}{6}$. After the first four terms, she saw the underlying patterns, and could continue the series for the next many terms.

So the definition of $\pi$, $\pi$ to 10 decimals and the Liebniz-Gregory and Euler series became the "passwords" that she had to write down, to gain access to the computer for her games. Being only 4, Kathryn had it easier — all she had to write down was the infinite series for "$e$". She was especially proud of the fact that she learnt the meaning of "factorial" (!) before her elder sister did.

The effortless encounters of Rebecca and Kathryn with the infinite series for $\pi$ and $e$, suggested to me the possibility that many people could also enjoy an easy introduction to the pleasures and beauty of mathematics the painless way. For far too long "mathematics" has been synonymous with "boredom", "pain", and "total incomprehension" for far too many people. If little children can appreciate and enjoy the patterns of infinite series for $\pi$ and $e$, so can everyone else.

Therefore I decided to write this book as a gift to my granddaughters so that as they grow up, and at the appropriate time, they may be initiated into the pleasures of "pure mathematics" — one of the most beautiful subjects in the world, according to the eminent former Cambridge mathematician, Professor G.H. Hardy.

This book consists of two short sections. Section I is largely a visual treat, a feast for the eyes, with equations which can be read and enjoyed by almost everyone who has heard of $\pi$, and know the simple integers and fractions. The series have been selected for their beauty, elegance and simplicity from the huge domain of mathematics. There are literally an infinite number of such series in mathematics to choose from. Section I is best read at one sitting, just as one would read a book of art, full of beautiful pictures, with short commentaries. That way, the beauty and mystery of the patterns and the rhythms of many of the equations will fill the reader, or perhaps for those seeing them for the first time, with a sense of awe and amazement. It should set them wondering why they had never encountered such beauty in the math that they had studied in school. Some of the series are so simple and elegant in their symmetry, that even Kathryn, a four-year-old child, could continue the series after the first few terms. If non-mathematicians and those who "hated math in school" could finish Section I, the effort in writing this book would have been worthwhile.

Section II of the book is largely a feast for the mind, and requires a bit more effort and some background in mathematics, preferably up to high school level. Here in the most systematic way possible, I have chosen the simplest proofs for the beautiful equations given in Section I, so that readers will be able to derive the beautiful equations themselves. Reading and enjoying Section I alone would be pleasure enough. But there is far greater pleasure and a deeper sense of intellectual satisfaction to know that, with an extra bit of effort and some patience, one could follow in the footsteps, and think the same thoughts, as some of the greatest mathematics geniuses in the world, such as Newton, Leibniz, and Euler.

Albert Einstein, the greatest 20th century physicist, said:

> Everything should be made as simple as possible
> But not simpler.

So Section II begins with the "Easy Proofs" — proofs so simple that Rebecca could do some of them. These are then followed by the "Less Easy Proofs" and finally, the "Not-So-Easy Proofs".

Section II of the book is best savoured slowly like fine wine, preferably not more than one or two proofs at a time — unless you are a mathematician, in which case, the proofs would be effortless and plain sailing for you.

John Keats, the young 19th century romantic poet, said:

A thing of beauty is a joy forever
Its loveliness increases
It will never pass into nothingness.

In this book, you will see that many of the terms in the equations of great beauty do "pass into nothingness" as the series tends to infinity!

The mathematical ideas, equations, formulas, proofs, etc. in this book are about 200–400 years old. Thousands of mathematicians had written about them over the centuries.

What is original in this book is the method of presentation of the math, with the primary purpose of illustrating the beauty of pure math via the vehicle of infinite series. After I had finished writing the book, I went through the books in the math section of Borders in Singapore and San Francisco, and Waterstone's, Blackwell's and Foyle's in London. I did not find a single book that presented math in the way that I had. My hope is that readers of the book (especially of Section I) will find it entertaining and pleasurable. If readers also find it educational (especially readers of Section II), it would be a happy bonus.

So enjoy, and take pleasure in the infinite series of $\pi$, $e$ and other interesting numbers.

# Acknowledgments

Many tributaries feed a river. An ancient Chinese proverb says: "Those who drink of the water should remember its sources, and be grateful."

It is a privilege for me to acknowledge my gratitude to all who had influenced, directly or indirectly, my intellectual development in life. Indeed, one's intellect is the sum (and occasionally the product) of all the inputs over the years; of lessons learnt in schools, universities and workplaces, as well as of ideas gleaned from books read and from friends, teachers, professors and colleagues.

I record here my gratitude to the many people whose ideas and values I have acquired, absorbed and assimilated over the years. They run into thousands. Mathematically speaking, they are finite in number; but humanly speaking, they are countless because memory fails, and often I cannot remember the sources or the names of those whom I have learnt from. Hence, my sincere apologies to those whose names are not recorded here.

I wish to thank all mathematicians since recorded history, including our ancient mathematician-forebears — the Chinese, Indians, Babylonians, Egyptians, Arabs, Greeks, etc. — for their contributions to this beautiful subject, mathematics. In this day of ubiquitous search engines, I have decided to replace the conventional bibliographic listing with an expression of gratitude to the many authors whose books on mathematics and related subjects, had given me countless hours of reading pleasure and education. They are too many to name. They include the following in alphabetical order:

Rouse Ball
John Barrow
Petr Beckman
E. T. Bell
John Casti
Richard Courant
Paul Davies
Keith Devlin
Richard Feynman
Martin Gardner
Margaret Gow
Brian Greene
John Gribbin
G. H. Hardy

Stephen Hawking
Michio Kaku
Robert and Ellen Kaplan
Eli Maor
Barry Mazur
Paul Nahin
Dan Pedoe
Roger Penrose
Martin Rees
Herbert Robbins
Marcus du Sautoy
Simon Singh
Ian Stewart

Entry of these names with any search engine will give the titles of a huge number of excellent books which I found to be of great value.

Those I know personally whom I thank include:

My Mathematics teachers:
    Many, including Miss Ong and Mr Gan (in secondary
        school);
    Many, including Mrs Lam and Prof Guha (in university).

My Chemistry teachers:
    Many (in school);
    Many (in universities) including:
    Emeritus Prof Kiang Ai Kim (University of Singapore);
    Emeritus Prof Peter Huang (University of Singapore),
        my M.Sc. supervisor;

Emeritus Prof Dudley H. Williams (Cambridge University),
  my Ph.D. supervisor;
the late Emeritus Prof Lord Todd (Cambridge University),
  Nobel Laureate and my Master at Christ's College;
Emeritus Prof Carl Djerassi (Stanford University),
  my post-doctoral research supervisor, and
Emeritus Prof Joshua Lederberg (Stanford University),
  Nobel Laureate and leader of our research team in
  Artificial Intelligence.

My special thanks go to the many friends directly involved with the book. They include:

Prof Lily Shih, for her reading and critique of the proofs;
Lim Sook Cheng and her excellent team at World Scientific
Publishing, and Peh Chin Hua and Peh Soh Ngoh at Shing
Lee Publishers, for helping in so many ways in the production of the book; Sam Chan for editorial improvements; and
Pauline Chia for helping in the typing and formatting the texts.

Special thanks also go to:

Prof Shih Choon Fong and
Prof Cham Tao Soon
for so kindly reading the proofs and writing the Foreword
to the book.

Last, but not least, To God be the Glory.

# Contents

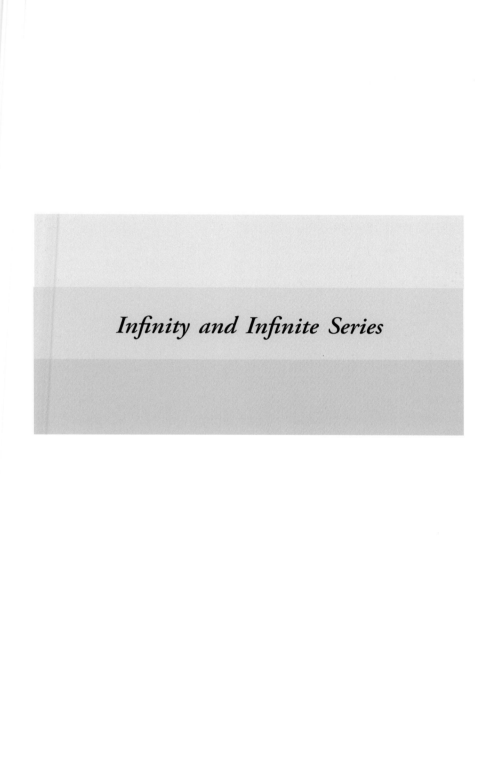

*Infinity and Infinite Series*

Numbers rule the Universe

**Pythagoras** (~580–500 BC)

❧

God is a geometer

**Plato** (~427–347 BC)

❧

God created everything by numbers

**Isaac Newton** (1642–1727)

❧

The Great Architect of the Universe
now begins to appear as a pure mathematician

**James Jean** (1877–1946)

❧

# The Sum of Integers

$$1 + 2 + 3 + 4 + 5 + 6 + 7 + \cdots$$

$$\rightarrow \infty$$

CHILDREN OFTEN ARGUE with one another over the number of objects that they have.

"I have 2 sweets",
"I have 3, one more than you".

Then they start escalating.
"I have 4",
"I have 10".
"I have 20",
"I have 100", and so on.

In next to no time, they begin to use bigger numbers, such as thousands, millions, and billions. Soon they run out of all the names of numbers that they know, and they move on to other subjects.

The concept of infinity is unknown to children and even to most adults. Indeed it was unknown to many earlier mathematicians who were confused by it, and avoided thinking or talking about the concept. This lasted until about the end of the 16th century.

Today mathematicians do not consider infinity (written as $\infty$) as a number. The sum of the endless addition of natural numbers does not add up to infinity. Rather, mathematicians say that the sum "tends to infinity", and signify it with the symbol "$\rightarrow \infty$". The dots at the end of the numbers (...) signify the endless continuation of terms.

It is easy to understand that if we keep adding up the integers endlessly, the sum gets larger and larger, and hence "tends to infinity".

---

(Mathematicians call natural whole numbers "integers". Once you start calling them integers, you can consider yourself a mathematician too!)

# The Sum of Factorials

$$1 + (1 \times 2) + (1 \times 2 \times 3) + (1 \times 2 \times 3 \times 4) + \cdots$$

$$\to \infty$$

$$1 + 2! + 3! + 4! + \cdots$$

$$\to \infty$$

IF THE SUM of integers tends to infinity, then it is obvious that the sum of the terms, which are the products of the consecutive integers from 1 up to the progressively higher integers indicated, also tends to infinity.

This is because while the first two terms are identical to those of the first series of integers, the third and subsequent terms are all greater than the corresponding terms of the first series of integers (e.g., $1 \times 2 \times 3$ is greater than 3). Mathematicians use the term "factorial", represented by (!) as an abbreviation for the product of the consecutive integers from 1 up to the integers indicated (e.g., $n!$ is $1 \times 2 \times 3 \times 4 \times \cdots \times n$).*

---

*Kathryn, 4, learnt the meaning of $n!$, and loves to recite it when asked.

# The Harmonic Series

$$\frac{1}{1} + \frac{1}{2} + \frac{1}{3} + \frac{1}{4} + \frac{1}{5} + \frac{1}{6} + \frac{1}{7} + \cdots$$

$$\rightarrow \infty$$

See Proof 14 → (page 165)

IF IT IS obvious that the sum of integers tends to infinity, what about the sum of their reciprocals (opposite page)?

First of all, let us note that the bigger the integer, the smaller is the reciprocal. So the individual reciprocals would get smaller and smaller, and tend to zero, as the integers themselves get larger and larger and tend to infinity.

Hence, intuitively we would expect the sum of reciprocals to tend to some constant, large though it may be.

Surprise, surprise! The sum of the reciprocals of integers also tends to infinity, even though the individual terms tend to zero.

Mathematicians refer to a series of terms whose sum tends to infinity as "divergent" — "a divergent series". Conversely, a series with a sum that tends to a fixed number (a constant), is referred to as "convergent" — "a convergent series".

The series (opposite page) enjoys a special place in mathematics and is given the name "the Harmonic Series" because the Greek mathematician Pythagoras and his followers believed that it was related to musical notes.

If the harmonic series is divergent, what about the sum of the same series, but with alternating signs: $\frac{1}{1} - \frac{1}{2} + \frac{1}{3} - \frac{1}{4} + \frac{1}{5} + \cdots$?

Also, what about the series with the reciprocals of the squares of the integers:

$$\frac{1}{1^2} + \frac{1}{2^2} + \frac{1}{3^2} + \frac{1}{4^2} + \frac{1}{5^2} + \cdots?$$

(This last problem had puzzled mathematicians for centuries, until it was solved in the 18th century.)

# The Two Halves of the Harmonic Series

$$\frac{1}{1} + \frac{1}{3} + \frac{1}{5} + \frac{1}{7} + \frac{1}{9} + \cdots$$

$$\to \infty$$

$$\frac{1}{2} + \frac{1}{4} + \frac{1}{6} + \frac{1}{8} + \frac{1}{10} + \cdots$$

$$\to \infty$$

See Proof 14 → (page 165)

IF THE SUM of the reciprocals of all integers tends to infinity, what about the sum of the series which consists of only some of the terms? What about the sum of the reciprocals of all the odd integers? Or the even integers?

Again, contrary to intuition, both series (opposite page) are divergent. Yes, the sums of both the series tend to infinity.

How many terms of the Harmonic Series do you need to remove before you get a convergent series? This is a problem that continues to puzzle mathematicians.

# The Geometric Series

$$\frac{1}{1} + \frac{1}{2} + \frac{1}{4} + \frac{1}{8} + \frac{1}{16} + \frac{1}{32} + \frac{1}{64} + \cdots$$
$$= 2$$

$$\frac{1}{2^0} + \frac{1}{2^1} + \frac{1}{2^2} + \frac{1}{2^3} + \frac{1}{2^4} + \frac{1}{2^5} + \frac{1}{2^6} + \cdots$$
$$= 2$$

See Proof 12 → (page 156)

HAVING SEEN THAT the sums of so many series of reciprocals tend to infinity, it is natural to ask the questions:

"Do the sums of all infinite series of reciprocals always tend to infinity?" Are there any infinite series of reciprocals which sum to a constant?"

As it turns out, there are numerous infinite series which do sum to constants.

"The Geometric Series", an infinite series of some of the reciprocals, sum to a small number, 2.

It is amazing how minor differences in the series can sometimes result in dramatic differences in their sums — 2 for "the Geometric Series", and "tending to infinity" for "the Harmonic Series" and their odd and even subsets.

"The Geometric Series" occupies an important place in the history of mathematics. It was known to the Greeks in the 5th century BC, but it created much confusion, and formed the basis of a famous series of problems known as "Zeno's Paradoxes".

# The Exponential Series

$$1 + \frac{1}{1} + \frac{1}{1 \times 2} + \frac{1}{1 \times 2 \times 3} + \frac{1}{1 \times 2 \times 3 \times 4} + \cdots$$

$$1 + \frac{1}{1!} + \frac{1}{2!} + \frac{1}{3!} + \frac{1}{4!} + \cdots$$

$$= 2.7182818284\ldots$$

$$= e$$

See Proof 22 → (page 181)

THE SUM OF integers and the sum of their reciprocals both tend to infinity. The sum of factorials tend to infinity. So would the sum of the reciprocals of factorials also tend to infinity?

Surprise, surprise! It doesn't! It sums to a small number (1.7182818284 ...). If we add 1 to the series, we get an extremely important series (opposite page), known as "the Exponential Series". This series was discovered in 1665 by Isaac Newton (1642–1727), the famous English mathematics genius, physicist, astronomer, theologian and polymath. Newton was working at home during a break from his formal studies at Cambridge University, which had to close as a result of a plague from 1665–67.

Another famous mathematician, Leonhard Euler (1707–1783) gave the symbol "$e$" to the sum of the series (opposite), which sums to 2.7182818284 ... . "$e$", always written in small letter, turned out to be an important universal constant, featuring in many branches of mathematics, science and technology, from the humble nautilus shell to the immense galaxies in the Universe.

# A Logarithmic Series

$$\frac{1}{1} - \frac{1}{2} + \frac{1}{3} - \frac{1}{4} + \frac{1}{5} - \frac{1}{6} + \frac{1}{7} - \cdots$$

$$= \log_{\text{natural}} 2$$

$$= \ln 2$$

$$= 0.6931471805\ldots$$

See Proof 15 → (page 167)

ENCOURAGED BY THE two series of reciprocals which sum to constants, one wonders if there are series with alternating terms (i.e. positive terms followed by negative terms, alternately) which sum to constants.

For example, the harmonic series, the sum of reciprocals of all the integers, tends to infinity. What about its sister series with alternating terms (opposite)?

This series turns out to be a special case of a class of logarithm series, and sums to the natural logarithm of 2 (which is 0.6931471805 …).

# The Liebniz-Gregory Series

$$\frac{1}{1} - \frac{1}{3} + \frac{1}{5} - \frac{1}{7} + \frac{1}{9} - \frac{1}{11} + \frac{1}{13} - \cdots$$

$$= \frac{\pi}{4}$$

$$= 0.78539816\ldots$$

See Proof 40 → (page 213)

ANOTHER SERIES WITH alternating terms is that of the reciprocals of odd integers (opposite page). Remember, the sum of the reciprocals of odd integers tends to infinity. This series of alternating terms is one of the most beautiful series involving $\pi$, and is called "the Liebniz-Gregory Series". It sums to a total of $\frac{\pi}{4}$, and hence serves to introduce us to the wonderful world of the constant $\pi$ — considered by most mathematicians as the most important constant in the world.

# The Definition of $\pi$

$$\pi = \frac{\text{Circumference of a circle}}{\text{Diameter}}$$

$$= \frac{C}{D}$$

$$= \frac{C}{2R}$$

THE EARLIEST RECORDS of mankind's awareness of $\pi$ are to be found among the Babylonians and Egyptians. Some four thousand years ago, they knew about $\pi$, as the ratio of the circumference of a circle to its diameter.

The Babylonians gave the value of $\pi$ as $\frac{25}{8}$, while the Egyptians used $\left(\frac{4}{3}\right)^4$, which works out to be $\frac{256}{81}$. It is amazing that the Babylonian value is only 0.5% off the correct value of $\pi$, while the Egyptian estimate is 0.6% off. Today, students in elementary schools routinely use the estimate of $\frac{22}{7}$ for $\pi$, which despite its simplicity is only 0.04% off its correct value. This estimate is attributed to the great Greek mathematician and physicist, Archimedes (287–212 BC). Yes, he's the one who ran naked in the streets and shouted "Eureka" after he discovered the "principle of displacement" of water. A more accurate approximation of $\pi$, but still a simple ratio is $\frac{335}{113}$. This gives $\pi$ accurate to 6 decimals.

$\pi$ is also essential for the calculation of a number of different properties of curved figures and objects:

Area of a circle = $\pi R^2$   ($R$ is the radius)
Surface area of a sphere = $4\pi R^2$
Volume of a sphere = $\frac{4}{3}\pi R^3$
Surface Area of a hollow cylinder = $2\pi RH$
    ($H$ is the height of the cylinder.)
Volume of a Cylinder = $\pi R^2 H$

This simple number, $\pi$, has proven to be an extremely important universal constant that finds application in many branches of mathematics, science and technology, beginning with the simple circle and sphere. At the other extreme of the complexity spectrum, $\pi$ also features in one of Albert Einstein's field equations for his Theory of General Relativity (1916) which described mathematically how the forces of gravity arise out of the curvature of the space-time continuum:

$$\left( R_{ab} - \frac{1}{2}Rg_{ab} = \frac{8\pi}{c^4}GT_{ab} \right)$$

# Divergent and Convergent Series

1. $1 + 2 + 3 + 4 + 5 + 6 + \cdots \qquad \rightarrow \infty$

2. $1! + 2! + 3! + 4! + 5! + 6! + \cdots \qquad \rightarrow \infty$

3. $\dfrac{1}{1} + \dfrac{1}{2} + \dfrac{1}{3} + \dfrac{1}{4} + \dfrac{1}{5} + \dfrac{1}{6} + \cdots \qquad \rightarrow \infty$

4. $\dfrac{1}{1} + \dfrac{1}{3} + \dfrac{1}{5} + \dfrac{1}{7} + \dfrac{1}{9} + \cdots \qquad \rightarrow \infty$

5. $\dfrac{1}{2} + \dfrac{1}{4} + \dfrac{1}{6} + \dfrac{1}{8} + \dfrac{1}{10} + \cdots \qquad \rightarrow \infty$

6. $\dfrac{1}{1} + \dfrac{1}{2} + \dfrac{1}{4} + \dfrac{1}{8} + \dfrac{1}{16} + \cdots \qquad = 2$

7. $\dfrac{1}{2^0} + \dfrac{1}{2^1} + \dfrac{1}{2^2} + \dfrac{1}{2^3} + \dfrac{1}{2^4} + \cdots \; = 2$

8. $1 + \dfrac{1}{1!} + \dfrac{1}{2!} + \dfrac{1}{3!} + \dfrac{1}{4!} + \cdots \qquad = e \; (2.71828\ldots)$

9. $\dfrac{1}{1} - \dfrac{1}{2} + \dfrac{1}{3} - \dfrac{1}{4} + \dfrac{1}{5} - \cdots \qquad = \ln 2 \; (0.69314\ldots)$

10. $\dfrac{1}{1} - \dfrac{1}{3} + \dfrac{1}{5} - \dfrac{1}{7} + \dfrac{1}{9} - \cdots \qquad = \dfrac{\pi}{4} \; (0.785139\ldots)$

π-series

Know you of this fair work?
Beyond the infinite and boundless ...

**William Shakespeare** (1564–1616)

§


Mathematics possesses not only truth,
but some supreme beauty

**Bertrand Russell** (1872–1970)

§

# The Liebniz-Gregory Series

$$\frac{1}{1} - \frac{1}{3} + \frac{1}{5} - \frac{1}{7} + \frac{1}{9} - \frac{1}{11} + \frac{1}{13} - \cdots$$

$$= \frac{\pi}{4}$$

See Proof 40 → (page 213)

LET US NOW make a more detailed acquaintance with the Liebniz-Gregory series, which we saw in our introduction to "convergent series". $\pi$ is just the simple ratio of the circumference of a circle to its diameter — a physical constant, measurable with a string. Yet in one of the most inexplicable results of human creativity, $\pi$ is expressible in an infinite number of equations. The following pages feature only some of the most beautiful, most elegant and simplest of such series involving $\pi$.

A Scottish mathematician, James Gregory (1638–1675) discovered the equation (opposite page) in 1671 at the age of 33, four years before his creative career was cut short by his premature demise. Independently, the great German mathematics genius Gottried Leibniz (1646–1716) also discovered the same equation in 1674. Hence this equation is often called the Leibniz-Gregory series. Unfortunately, the sum tends to the correct value of $\pi$ extremely slowly — some 600 terms are required to give the value of $\pi$ correct to 2 decimal places! (For comparison, the simple ratio, $\frac{22}{7}$, also gives $\pi$ correct to 2 decimal places!)

The Liebniz-Gregory series illustrates well the truism, that the most beautiful things in the world may not necessarily be the most useful.

# "Grampa's Series"

$$\frac{1}{1\cdot 3} + \frac{1}{5\cdot 7} + \frac{1}{9\cdot 11} + \frac{1}{13\cdot 15} + \frac{1}{17\cdot 19} + \cdots$$

$$= \frac{\pi}{2\cdot 4}$$

See Proof 4 → (page 148)

THIS SERIES IS a late addition to the book. I had finished writing the book and was playing around with infinite series which sum to interesting numbers, such as

$$\frac{1}{1 \cdot 2} + \frac{1}{2 \cdot 3} + \frac{1}{3 \cdot 4} \cdots$$

$$\frac{1}{1 \cdot 2} + \frac{2}{3 \cdot 4} + \frac{3}{5 \cdot 6} \cdots \text{etc.}$$

Even as I was playing with the different series, I chanced upon the series with odd integers:

$$\frac{1}{1 \cdot 3} + \frac{1}{5 \cdot 7} + \frac{1}{9 \cdot 11} + \frac{1}{13 \cdot 15} \cdots$$

The sum of the series worked out to be half the Liebniz-Gregory Series, which sums to $\frac{\pi}{4}$. What an "eureka" moment. What serendipity! And the proof is simple.

Obviously since the Liebniz-Gregory equation was discovered some three hundred years ago, thousands (probably millions) of mathematicians must have discovered this equation too. It's just that I have not seen it before in the mathematics literature that I've read so far.

This incident shows that there are still "eureka" moments when one discovers something by one's own effort. It does not matter, as in the case of Liebniz, that someone else had discovered it before.

I told my granddaughter Rebecca after the discovery. Now she calls the series "Grampa's Series", and writes it out faithfully as another of her "passwords".

# The Euler Series

$$\frac{1}{1^2} + \frac{1}{2^2} + \frac{1}{3^2} + \frac{1}{4^2} + \frac{1}{5^2} + \frac{1}{6^2} + \cdots$$

$$= \frac{\pi^2}{6}$$

See Proof 41 → (page 214)

IN THE 17TH and 18th century, mathematicians were seeking ways of summing infinite series of all sorts and patterns. Among the most intriguing of series is the sum of the reciprocals of the squares of all the positive integers. An extremely elegant and deceptively simple series, the sum eluded the efforts of all the mathematicians, including such geniuses as Newton, Liebniz and Jakob Bernoulli (1654–1705) and his famous family of outstanding mathematicians. The lack of a solution was definitely not for want of trying.

Then came Euler.

Leonhard Euler (1707–83), a Swiss mathematician, was to prove himself to be the genius of geniuses, one of the greatest mathematicians of all time. In 1734, at the tender age of 27, Euler derived this equation for $\frac{\pi^2}{6}$ and ended the search by generations of mathematicians. In the process, Euler established his reputation as a "Wizard Mathematician", and went on over the next half a century of active creativity, to contribute to so many branches of mathematics and physics that his name Euler is attached to so many equations, formulas, constants and series in mathematics and science that one has to specify the field before one could be sure which Euler equation or formula one is discussing.

In the later parts of this book, we shall be walking in his footsteps and thinking his thoughts, some 250 years after him. What a privilege for us!

# More Euler Series

$$\frac{1}{1^2} + \frac{1}{3^2} + \frac{1}{5^2} + \frac{1}{7^2} + \frac{1}{9^2} + \cdots$$

$$= \frac{\pi^2}{8}$$

$$\frac{1}{2^2} + \frac{1}{4^2} + \frac{1}{6^2} + \frac{1}{8^2} + \frac{1}{10^2} + \cdots$$

$$= \frac{\pi^2}{24}$$

$$\frac{1}{1^2} - \frac{1}{2^2} + \frac{1}{3^2} - \frac{1}{4^2} + \frac{1}{5^2} - \frac{1}{6^2} + \cdots$$

$$= \frac{\pi^2}{12}$$

See Proof 42 → (page 216)

AFTER EULER'S STUPENDOUS ground-breaking achievement with $\frac{\pi^2}{6}$, similar beautiful gems rained from the sky with relative ease. By simple arithmetical rearrangements of the terms in $\frac{\pi^2}{6}$ (the series is absolutely convergent, so we can rearrange the terms without erroneous results — see "A Note of Caution"; page 175), three more Euler series can be obtained, all three as beautiful as the original. The series of alternating terms (plus followed by minus) sums to $\frac{\pi^2}{12}$. The series of reciprocals of odd integers sum to $\frac{\pi^2}{8}$, while those of even integers sum to $\frac{\pi^2}{24}$.

What a family of beauties for pure mathematics!

# Vieté's Equation

$$\frac{\sqrt{2}}{2} \cdot \frac{\sqrt{2+\sqrt{2}}}{2} \cdot \frac{\sqrt{2+\sqrt{2+\sqrt{2}}}}{2} \dots$$

$$= \frac{2}{\pi}$$

See Proof 43 → (page 217)

WHILE MATHEMATICS GENIUSES such as Newton, Liebniz and Euler deserve to be acclaimed for their voluminous ground-breaking achievements, we should not forget the signal contributions of some of their predecessors in mathematics.

Francois Vieté (1540–1603) was a French lawyer, politician, diplomat and amateur mathematician. One of his many mathematical contributions is the expression of $\pi$ as an infinite product involving only the integer 2, in an infinite nesting of square roots. Vieté's expression marked a milestone in the history of mathematics. It was the first equation incorporating the concept of an infinite process, even though it was not explicitly spelt out as such. The dots (...) in the equation denotes continuing the process indefinitely.

If we think that the expression looks difficult, remember Vieté published it in 1593, more than four hundred years ago!

# Wallis' Equation

$$\frac{2 \cdot 2}{1 \cdot 3} \cdot \frac{4 \cdot 4}{3 \cdot 5} \cdot \frac{6 \cdot 6}{5 \cdot 7} \cdot \frac{8 \cdot 8}{7 \cdot 9} \cdots$$

$$= \frac{\pi}{2}$$

See Proof 44 → (page 219)

JOHN WALLIS (1616–1702), an Englishman, was the second mathematician (after Vieté) to give an equation for $\pi$ in terms of the product of an infinite number of terms. His equation, published in 1655, was derived by laborious calculations, without the benefit of later developments in trigonometry and calculus in subsequent centuries. Wallis was also the first mathematician to use the symbol $\infty$ to represent infinity.

# More Wallis Equations

$$\frac{3\cdot 3}{2\cdot 4}\cdot\frac{6\cdot 6}{5\cdot 7}\cdot\frac{9\cdot 9}{8\cdot 10}\cdots$$

$$=\frac{2\pi}{3\sqrt{3}}$$

$$\frac{4\cdot 4}{3\cdot 5}\cdot\frac{8\cdot 8}{7\cdot 9}\cdot\frac{12\cdot 12}{11\cdot 13}\cdots$$

$$=\frac{\pi}{2\sqrt{2}}$$

$$\frac{6\cdot 6}{5\cdot 7}\cdot\frac{12\cdot 12}{11\cdot 13}\cdot\frac{18\cdot 18}{17\cdot 19}\cdots$$

$$=\frac{\pi}{3}$$

See Proof 45 → (page 220)

THE MODERN DERIVATION of Wallis' equation makes use of the function $\frac{\sin x}{x}$. By using different values for $x$, other Wallis' equations summing to $\frac{\pi}{3}$, $\frac{\pi}{2\sqrt{2}}$ and $\frac{2\pi}{3\sqrt{3}}$ follow.

Again another family of beauties; maybe not as elegant as Euler's!

$$1 - \frac{1}{3}\left(\frac{1}{3^1}\right) + \frac{1}{5}\left(\frac{1}{3^2}\right) - \frac{1}{7}\left(\frac{1}{3^3}\right) + \frac{1}{9}\left(\frac{1}{3^4}\right) - \frac{1}{11}\left(\frac{1}{3^5}\right) + \cdots$$

$$= \frac{\pi}{2\sqrt{3}}$$

See Proof 46 → (page 222)

THE BRANCH OF mathematics known as trigonometry gives rise to a number of infinite series for the common functions of sine, and cosine, and inverse functions, arcsin and arctan. Using radians (where the angle around a point is treated as $2\pi$ instead of $360°$), these equations give numerous infinite series for $\pi$.

The famous Liebniz-Gregory equation for $\frac{\pi}{4}$ is the best known example of the arctan series. The equation on the opposite page is another sister series from the arctan function. Note especially the pattern of alternating positive and negative terms, and the reciprocals of odd integers — characteristic of the arctan family of infinite series. A sister from the same parent it may be, yet one is truly beautiful while the other, less so.

Not quite the beauty and the beast though.

# Euler's Formula

$$\arctan\left(\frac{1}{n}\right) = \arctan\left(\frac{1}{n+1}\right) + \arctan\left(\frac{1}{n(n+1)+1}\right)$$

$$\arctan\left(\frac{1}{1}\right) = \arctan\left(\frac{1}{2}\right) + \arctan\left(\frac{1}{3}\right)$$

See Proof 47 → (page 223)

YET ANOTHER WAY of expressing the arctan function (opposite), this time in terms of the sum of two other arctan functions.

This equation is known as Euler's formula, first published in 1738.

Note that the equation involves the variable $n$ (any integer). This means that by using different values for $n$ (i.e., $n = 1, 2, 3, \ldots$), we can derive an infinite number of equations (of course, not all of them are simple, beautiful and elegant). Using $n = 1$, the general Euler formula gives the beautiful second equation. Nothing can be simpler than 1, 2 and 3!

("What has this formula got to do with $\pi$?" you may well ask — turn the page.)

# Machin's Equation

$$4 \arctan\left(\frac{1}{5}\right) - \arctan\left(\frac{1}{239}\right)$$

$$= \frac{\pi}{4}$$

See Proof 48 → (page 224)

RETURNING TO THE often quoted beautiful Liebniz-Gregory expression for $\frac{\pi}{4}$, we remember how slowly it sums for the calculation of $\pi$. Here is another expression for $\frac{\pi}{4}$ (opposite) — another arctan formula which comes to the rescue of Liebniz-Gregory.

This less elegant expression is called the "Machin Equation", named in honour of John Machin (1680–1752), an English mathematician and astronomer. Using the Euler formula which gave the beautiful 1-2-3 arctan formula, Machin derived this workhorse equation (opposite). This equation has the merit of usefulness over beauty. It gives a very accurate and rapid approximation for $\pi$. Using only two terms in this infinite arctan series, the value of $\pi$ can be obtained, accurate to two decimal points (equivalent to some 600 terms in the Liebniz-Gregory equation). With this formula, Machin calculated the value of $\pi$ to 100 decimal points in 1706, a truly memorable milestone indeed in the history of $\pi$.

Since then, new "Machin-type expressions" using more complex arctan expressions have been derived by $\pi$-experts. A team of Japanese mathematicians at the Tokyo University, led by Yasumas Kanada currently (Sept 2002), holds the world record for calculating the value of $\pi$ to 1.241 trillion digits (yes, 1,241,000,000,000 digits long).

$$1 + \frac{1}{3 \cdot 2^3} + \frac{3}{4} \cdot \frac{1}{5 \cdot 2^5} + \frac{3 \cdot 5}{4 \cdot 6} \cdot \frac{1}{7 \cdot 2^7} + \cdots$$

$$= \frac{\pi}{3}$$

See Proof 49 → (page 226)

EULER'S GREAT ACHIEVEMENT for $\frac{\pi^2}{6}$ used a complex equation involving the function $\frac{\sin x}{x}$. This function, in its different manifestations, or after different mathematical manipulations, gives the Euler $\frac{\pi^2}{8}$, $\frac{\pi^2}{12}$ and $\frac{\pi^2}{24}$ series, and the famous Wallis equations for $\pi$, with

$$\frac{2\cdot 2}{1\cdot 3}\cdots, \frac{3\cdot 3}{2\cdot 4}\cdots, \frac{4\cdot 4}{3\cdot 5}\cdots \text{ and } \frac{6\cdot 6}{5\cdot 7}\cdots$$

infinite product series.

A related function arcsin $x$ can give yet another family of expressions for $\pi$, though these are less elegant and more complicated. One such expression is given (opposite) for the purpose of illustration.

Not a beauty by a long shot.

$$\frac{1}{1}\left(\frac{1}{2}+\frac{1}{3}\right)-\frac{1}{3}\left(\frac{1}{2^3}+\frac{1}{3^3}\right)+\frac{1}{5}\left(\frac{1}{2^5}+\frac{1}{3^5}\right)-\cdots$$

$$=\frac{\pi}{4}$$

See Proof 50 → (page 228)

THE EXPRESSIONS FOR arctan functions play a disproportionate role in the derivation of beautiful infinite series for $\pi$. The equation (opposite) is superficially similar to the beautiful Liebniz-Gregory expression with alternating reciprocals of odd integers, but embedded within each term (in brackets) is another sum with some degree of complexity.

Yet, surprise, surprise, the sum of infinite terms of both equations are identical — $\frac{\pi}{4}$!

What beauty! How amazing mathematics can be!

And yes, this complex expression uses the arctan series too, but in a more convoluted manner.

$$\frac{1}{4}\tan\left(\frac{\pi}{4}\right) + \frac{1}{8}\tan\left(\frac{\pi}{8}\right) + \frac{1}{16}\tan\left(\frac{\pi}{16}\right) + \cdots$$

$$= \frac{1}{\pi}$$

$$\frac{1}{2^2}\tan\left(\frac{\pi}{2^2}\right) + \frac{1}{2^3}\tan\left(\frac{\pi}{2^3}\right) + \frac{1}{2^4}\tan\left(\frac{\pi}{2^4}\right) + \cdots$$

$$= \frac{1}{\pi}$$

See Proof 51 → (page 229)

THERE IS LITERALLY an infinite number of infinite series for $\pi$ expressions, because many of these equations can be written in terms of the general integer, $n$. By using different values for $n$, different infinite expressions for $\pi$ are obtained, with different degrees of beauty, elegance, simplicity and usefulness.

Only a small handful of those expressions, considered among the more beautiful, has been chosen for this book.

To end this chapter on $\pi$ expressions, let's conclude with an infinite expression that is different from all the preceding expressions! This final $\pi$ equation is now expressed in infinite terms involving $\pi$ itself. The reason why this expression is used to close the chapter on $\pi$ is its sheer beauty.

And, of course, it was first derived by the genius of geniuses, the "Wizard Mathematician" Leonhard Euler, some three hundred years ago!

# $\pi$-Series

1. $\dfrac{1}{1} - \dfrac{1}{3} + \dfrac{1}{5} - \dfrac{1}{7} + \dfrac{1}{9} - \cdots \qquad = \dfrac{\pi}{4}$

2. $\dfrac{1}{1\cdot3} + \dfrac{1}{5\cdot7} + \dfrac{1}{9\cdot11} + \dfrac{1}{13\cdot15} + \cdots \quad = \dfrac{\pi}{2\cdot4}$

3. $\dfrac{1}{1^2} + \dfrac{1}{2^2} + \dfrac{1}{3^2} + \dfrac{1}{4^2} + \cdots \qquad = \dfrac{\pi^2}{6}$

4. $\dfrac{1}{1^2} + \dfrac{1}{3^2} + \dfrac{1}{5^2} + \dfrac{1}{7^2} + \cdots \qquad = \dfrac{\pi^2}{8}$

5. $\dfrac{1}{2^2} + \dfrac{1}{4^2} + \dfrac{1}{6^2} + \dfrac{1}{8^2} + \cdots \qquad = \dfrac{\pi^2}{24}$

6. $\dfrac{1}{1^2} - \dfrac{1}{2^2} + \dfrac{1}{3^2} - \dfrac{1}{4^2} + \cdots \qquad = \dfrac{\pi^2}{12}$

7. $\dfrac{\sqrt{2}}{2} \cdot \dfrac{\sqrt{2+\sqrt{2}}}{2} \cdot \dfrac{\sqrt{2+\sqrt{2+\sqrt{2}}}}{2} \cdots = \dfrac{2}{\pi}$

8. $\dfrac{2\cdot2}{1\cdot3} \cdot \dfrac{4\cdot4}{3\cdot5} \cdot \dfrac{6\cdot6}{5\cdot7} \cdot \dfrac{8\cdot8}{7\cdot9} \cdots \qquad = \dfrac{\pi}{2}$

9. $\dfrac{3\cdot3}{2\cdot4} \cdot \dfrac{6\cdot6}{5\cdot7} \cdot \dfrac{9\cdot9}{8\cdot10} \cdot \dfrac{12\cdot12}{11\cdot13} \cdots \qquad = \dfrac{2\pi}{3\sqrt{3}}$

10. $\dfrac{4\cdot4}{3\cdot5} \cdot \dfrac{8\cdot8}{7\cdot9} \cdot \dfrac{12\cdot12}{11\cdot13} \cdot \dfrac{16\cdot16}{15\cdot17} \cdots \qquad = \dfrac{\pi}{2\sqrt{2}}$

11. $\dfrac{6 \cdot 6}{5 \cdot 7} \cdot \dfrac{12 \cdot 12}{11 \cdot 13} \cdot \dfrac{18 \cdot 18}{17 \cdot 19} \cdot \dfrac{24 \cdot 24}{23 \cdot 25} \cdots \qquad = \dfrac{\pi}{3}$

12. $1 - \dfrac{1}{3}\left(\dfrac{1}{3^1}\right) + \dfrac{1}{5}\left(\dfrac{1}{3^2}\right) - \dfrac{1}{7}\left(\dfrac{1}{3^3}\right) + \cdots \qquad = \dfrac{\pi}{2\sqrt{3}}$

13. $\arctan\left(\dfrac{1}{n}\right) = \arctan\left(\dfrac{1}{n+1}\right) + \arctan\left(\dfrac{1}{n(n+1)+1}\right)$

14. $\arctan\left(\dfrac{1}{1}\right) = \arctan\left(\dfrac{1}{2}\right) + \arctan\left(\dfrac{1}{3}\right)$

15. $4\arctan\left(\dfrac{1}{5}\right) - \arctan\left(\dfrac{1}{239}\right) \qquad = \dfrac{\pi}{4}$

16. $1 + \dfrac{1}{3 \cdot 2^3} + \dfrac{3}{4} \cdot \dfrac{1}{5 \cdot 2^5} + \dfrac{3 \cdot 5}{4 \cdot 6} \cdot \dfrac{1}{7 \cdot 2^7} + \cdots \qquad = \dfrac{\pi}{3}$

17. $\dfrac{1}{1}\left(\dfrac{1}{2} + \dfrac{1}{3}\right) - \dfrac{1}{3}\left(\dfrac{1}{2^3} + \dfrac{1}{3^3}\right) + \dfrac{1}{5}\left(\dfrac{1}{2^5} + \dfrac{1}{3^5}\right) - \cdots \quad = \dfrac{\pi}{4}$

18. $\dfrac{1}{4}\tan\left(\dfrac{\pi}{4}\right) + \dfrac{1}{8}\tan\left(\dfrac{\pi}{8}\right) + \dfrac{1}{16}\tan\left(\dfrac{\pi}{16}\right) + \cdots \qquad = \dfrac{1}{\pi}$

19. $\dfrac{1}{2^2}\tan\left(\dfrac{\pi}{2^2}\right) + \dfrac{1}{2^3}\tan\left(\dfrac{\pi}{2^3}\right) + \dfrac{1}{2^4}\tan\left(\dfrac{\pi}{2^4}\right) + \ldots = \dfrac{1}{\pi}$

*e-series*

To ... hold infinity
in the palm of your hand

**William Blake** (1757–1827)

§


The mathematician's patterns
like those of the painter's or the poet's
must be beautiful
the ideas
like the colours or the words
must fit together in a harmonious way
Beauty is the first test
There is no permanent place in the world
for ugly mathematics

**G H Hardy** (1877–1947)

§

# The Definition of $e^x$

If $n \rightarrow \infty$

$$\left(1 + \frac{x}{n}\right)^n = 1 + \frac{x}{1!} + \frac{x^2}{2!} + \frac{x^3}{3!} + \cdots$$

$$= e^x$$

ANYONE WHO HAS deposited money in a bank will know about the concept of compound interest, where the amount of interest paid with each passing year increases as the capital grows from the accumulation of interests from the preceding years. The equation for calculating such compound interest is $(1+\frac{1}{m})^n$, where 1 is the capital, $\frac{1}{m}$ is the annual interest, and $n$ is the number of years that the money has been deposited. (If the interest is 5%, then $m = 20$)

This simple formula is similar to the more general term $(1+\frac{x}{n})^n$, which can be expressed as a beautiful infinite series when $n \to \infty$. In mathematics, the sum of this series is given the simple notation, $e^x$, e being an abbreviation for "exponential".

e (e is always in small letter) was first used by Leonhard Euler, and is referred to as Euler's constant. $e^x$ has the unique distinction of being the only function in mathematics to have the same derivative as itself. (In graphical terms, the function $e^x$ has the same value as the tangent at the same point.)

$$e^x = 1 + \frac{x}{1!} + \frac{x^2}{2!} + \frac{x^3}{3!} + \frac{x^4}{4!} + \cdots$$

$$e^{-x} = 1 - \frac{x}{1!} + \frac{x^2}{2!} - \frac{x^3}{3!} + \frac{x^4}{4!} - \cdots$$

BY REPLACING $x$ with $(-x)$ and observing that $(-x)$ raised to odd powers gives negative terms, and $(-x)$ raised to even powers gives positive terms, we derive the infinite series for $e^{-x}$. This is the sister equation for $e^x$, with alternating positive and negative terms.

$$e^x = 1 + \frac{x}{1!} + \frac{x^2}{2!} + \frac{x^3}{3!} + \frac{x^4}{4!} + \cdots$$

$$e^1 = e = 1 + \frac{1}{1!} + \frac{1}{2!} + \frac{1}{3!} + \frac{1}{4!} + \cdots$$

$$= 2.7182818284\ldots$$

$$e^{-1} = \frac{1}{e} = 1 - \frac{1}{1!} + \frac{1}{2!} - \frac{1}{3!} + \frac{1}{4!} - \cdots$$

$$= 0.3678794411\ldots$$

$$e = 1 + 1\left(1 + \frac{1}{2}\left(1 + \frac{1}{3}\left(1 + \frac{1}{4}(1 + \ldots)\right)\right)\right)$$

See Proofs 22, 23 → (pages 181, 182)

By ASSIGNING $x = 1$, we derive yet another beautiful infinite series, "the exponential series", which we met earlier on in our introduction to the convergent series. This series forms the basis of many other beautiful series which we will see in the following pages. Most amazing of all, e turns out to be a universal constant that is present in many aspects of life, from the small sunflower to the gigantic spiral galaxies in the universe. (A simple fraction with some degree of symmetry that gives e accurate to four decimal points is $\frac{878}{323}$.)

e has been calculated up to 17 billion decimal points by X. Gourdon and S. Kondo in the year 2000.

Interestingly, e can also be expressed as an infinite nesting of great symmetry, involving all the integers (opposite page).

$$\frac{1}{1!} + \frac{2}{2!} + \frac{3}{3!} + \frac{4}{4!} + \frac{5}{5!} + \frac{6}{6!} + \cdots$$

$$= e$$

See Proof 24 → (page 185)

WHAT A DELECTABLE way to begin our exploration of the many e series.

Would you believe that the series (opposite page) also sums to e?

Would you also believe that there is a simple one-step proof for it?

$$1 + \frac{3}{2!} + \frac{5}{4!} + \frac{7}{6!} + \frac{9}{8!} + \frac{11}{10!} + \cdots$$

$$= e$$

See Proof 25 → (page 186)

ANOTHER BEAUTIFUL INFINITE series for e! Amazing isn't it, the number of different infinite series that we can express the universal constant in. Note that this series has even integer factorials only!

A simple sum, with a simple proof.

$$\frac{1}{1!} + \frac{2}{3!} + \frac{3}{5!} + \frac{4}{7!} + \frac{5}{9!} + \frac{6}{11!} + \cdots$$

$$= \frac{1}{2}e$$

See Proof 26 → (page 188)

A SERIES THAT looks very similar to the previous series, but this time with odd integers factorials only. Interestingly, it sums to only $\frac{1}{2}$ e. Another beautiful series, with a simple proof.

$$\frac{2}{3!} + \frac{4}{5!} + \frac{6}{7!} + \frac{8}{9!} + \frac{10}{11!} + \frac{12}{13!} + \cdots$$

$$= e^{-1}$$

$$= \frac{1}{e}$$

See Proof 27 → (page 189)

ANOTHER VERY SIMILAR series, still with odd integer factorials only, but now with even integers in the numerators. Isn't $e^{-1}$ ($\frac{1}{e}$) made up of alternating positive and negative terms? What happened to the negative terms? And the even integer factorials?

Do the individual terms look like fractions that seem to be approximating to 1? Why, then, doesn't the sum tend to infinity, but to a small sum $e^{-1}$ (0.3678 ...)?

$$\frac{1}{2!} + \frac{2}{3!} + \frac{3}{4!} + \frac{4}{5!} + \frac{5}{6!} + \frac{6}{7!} + \cdots$$

$$= 1$$

See Proof 28 → (page 190)

An even more interesting and beautiful series which sums to 1!

Why does this series with factorials in the denominators not sum to some constant involving e, like all the other series with factorials in their denominators?

$$1 + \frac{2}{1!} + \frac{3}{2!} + \frac{4}{3!} + \frac{5}{4!} + \frac{6}{5!} + \cdots$$

$$= 2e$$

See Proof 29 → (page 192)

WOULD YOU BELIEVE that the sum for this series is only 2e? It is so counter-intuitive because 2e is simply

$$2\left(1+\frac{1}{1!}+\frac{1}{2!}+\frac{1}{3!}+\frac{1}{4!}\cdots\right) \text{giving}$$

$$2+\frac{2}{1!}+\frac{2}{2!}+\frac{2}{3!}+\frac{2}{4!}\cdots$$

Intuitively, since the numerators get larger and larger as $n$ tends to infinity, one would expect the sum to be divergent and tend to infinity. Still, there is a simple proof that the sum is indeed a small multiple of e.

$$\frac{1^2}{1!} + \frac{2^2}{2!} + \frac{3^2}{3!} + \frac{4^2}{4!} + \frac{5^2}{5!} + \frac{6^2}{6!} + \cdots$$

$$= 2e$$

See Proof 30 → (page 193)

How CAN A series involving squares (of integers) in the numerators sum to the same small constant (2e) just as in the previous simpler equation without squares in the numerators sums to 2e?

Don't the squares in the numerators get progressively larger and larger as $n$ tends to infinity? Why doesn't the series become divergent? How can it sum to 2e only?

$$\frac{1^2}{2!} + \frac{2^2}{3!} + \frac{3^2}{4!} + \frac{4^2}{5!} + \frac{5^2}{6!} + \frac{6^2}{7!} + \cdots$$

$$= (e-1)$$

See Proof 31 → (page 194)

KEEP YOUR EYES on the numerators and the denominators, and see them slide back and forth.

Amazing, a sum of only (e − 1) i.e., 1.7828 ... for an infinite series involving the squares of integers in the numerators!

$$1 + \frac{2^2}{1!} + \frac{3^2}{2!} + \frac{4^2}{3!} + \frac{5^2}{4!} + \frac{6^2}{5!} + \cdots$$

$$= 5e$$

See Proof 32 → (page 195)

HAVE WE SEEN this series before? Can you see how the numerators and denominators are sliding?

Only 5e for the sum?

$$\frac{1 \cdot 3}{2!} + \frac{2 \cdot 4}{3!} + \frac{3 \cdot 5}{4!} + \frac{4 \cdot 6}{5!} + \frac{5 \cdot 7}{6!} + \frac{6 \cdot 8}{7!} + \cdots$$

$$= (e + 1)$$

See Proof 33 → (page 196)

LET US NOW move on and play with more complex functions, this time involving products of integers in the numerators.

Increasing complexity, while still retaining simplicity in its sum.

$$\frac{1 \cdot 3}{2!} + \frac{3 \cdot 5}{4!} + \frac{5 \cdot 7}{6!} + \frac{7 \cdot 9}{8!} + \frac{9 \cdot 11}{10!} + \frac{11 \cdot 13}{12!} + \cdots$$

$$= \frac{e^2 + 2e - 1}{2e}$$

See Proof 34 → (page 197)

DOESN'T THIS SERIES look very similar to the previous equation? Why then is the sum now so complex, instead of the simple sums that we are so familiar with so far?

$$\frac{2}{1!3} - \frac{3}{2!4} + \frac{4}{3!5} - \frac{5}{4!6} + \frac{6}{5!7} - \frac{7}{6!8} + \cdots$$

$$= 3\left(\frac{1}{2} - \frac{1}{e}\right)$$

See Proof 35 → (page 199)

A SLIGHT SWITCH in the numerators and denominators, now with product functions in the denominators! And to add to the complexity, alternating positive and negative terms.

But the sum has become simpler once again! Don't be misled by the simplicity of the sum. The proof is really lengthy and complicated.

$$\frac{1}{1!} + \frac{1+2}{2!} + \frac{1+2+3}{3!} + \frac{1+2+3+4}{4!} + \cdots$$

$$= \frac{3e}{2}$$

See Proof 36 → (page 201)

AN EVEN MORE complex function in the numerators, this time with a function involving the sum of large numbers of integers as $n \to \infty$!

Again more complex terms summing to a simple total.

$$\frac{1(2^2+1)}{2!} + \frac{2(3^2+1)}{3!} + \frac{3(4^2+1)}{4!} + \frac{4(5^2+1)}{5!} + \cdots$$

$$= (3e+1)$$

$$\frac{1(2^2)}{2!} + \frac{2(3^2)}{3!} + \frac{3(4^2)}{4!} + \frac{4(5^2)}{5!} + \cdots$$

$$= (3e)$$

$$\frac{1(2^2-1)}{2!} + \frac{2(3^2-1)}{3!} + \frac{3(4^2-1)}{4!} + \frac{4(5^2-1)}{5!} + \cdots$$

$$= (3e-1)$$

See Proof 37 → (page 202)

EVER INCREASING COMPLEXITY with more heavyweights in the numerators Still not tending to infinity, but summing to a simple total.

What about its sister series (3rd equation) with a negative sign in the bracket in the numerators? Again summing to a simple total, now a symmetrical image of the first equation.

And would you believe it — the middle equation, without the "1" in the bracket in the numerators — sums to the total without the "1".

What beautiful symmetry! Incredible, isn't it?

Can you play around with the numerators and denominators and create your own series? The e-series is extremely flexible like magic putty, and there are many more such series that you can create yourself by increasing, for example, the powers of the numerators, having simple functions in terms of $n$ in the numerators, having only even or odd factorials in the denominators, or even alternating the terms with positive and negative terms.

The methods involving summations (as shown in the proofs in Section II) are general and versatile, and can be applied to different series. Have fun!

$$\frac{e^{\frac{1}{1}}}{e^{\frac{1}{2}}} \cdot \frac{e^{\frac{1}{3}}}{e^{\frac{1}{4}}} \cdot \frac{e^{\frac{1}{5}}}{e^{\frac{1}{6}}} \cdot \frac{e^{\frac{1}{7}}}{e^{\frac{1}{8}}} \cdot \frac{e^{\frac{1}{9}}}{e^{\frac{1}{10}}} \cdot \frac{e^{\frac{1}{11}}}{e^{\frac{1}{12}}} \cdots$$

$$= 2$$

See Proof 38 → (page 203)

ANOTHER EULER BEAUTY to please the eye, with all the $e$'s staring back at us! How can an integer as simple as 2 be expressed in so complex a manner as the quotient of two infinite products of fractional powers of the irrational universal constant $e$? And even the proof is simple!

But it took a genius like Euler to come up with it.

(Can you prove it yourself, and be an Euler, too?)

# "The Most Beautiful Equation in the World"
## Euler's Identity

$$e^{i\pi} = -1$$

$$e^{i\pi} + 1 = 0$$

See Proof 39 → (page 204)

WE END THE chapters on $\pi$ and e with an equation that has been acclaimed by numerous mathematicians as "The Most Beautiful Equation in the World".

With the five most important symbols in mathematics, namely e, $i$, $\pi$, 0, and 1, this simple equation links together the many branches of mathematics, including numerical analysis, geometry, trigonometry and complex numbers ($i$ is the square root of $(-1)$, and is called an imaginary number; it is also the parent of complex numbers which are expressed in terms of an imaginary part and a real part.)

Yes, this beautiful equation was first discovered by none other than Leonhard Euler! And it is known as "Euler's Identity".

# e-Series

1. $1 + \dfrac{x}{1!} + \dfrac{x^2}{2!} + \dfrac{x^3}{3!} + \dfrac{x^4}{4!} + \cdots$    $= e^x$

2. $1 - \dfrac{x}{1!} + \dfrac{x^2}{2!} - \dfrac{x^3}{3!} + \dfrac{x^4}{4!} - \cdots$    $= e^{-x}$

3. $1 + \dfrac{1}{1!} + \dfrac{1}{2!} + \dfrac{1}{3!} + \dfrac{1}{4!} + \cdots$    $= e$

4. $1 - \dfrac{1}{1!} + \dfrac{1}{2!} - \dfrac{1}{3!} + \dfrac{1}{4!} - \cdots$    $= e^{-1}$

5. $1 + 1\left(1 + \dfrac{1}{2}\left(1 + \dfrac{1}{3}\left(1 + \dfrac{1}{4}(\cdots)\right)\right)\right)$    $= e$

6. $\dfrac{1}{1!} + \dfrac{2}{2!} + \dfrac{3}{3!} + \dfrac{4}{4!} + \dfrac{5}{5!} + \cdots$    $= e$

7. $1 + \dfrac{3}{2!} + \dfrac{5}{4!} + \dfrac{7}{6!} + \dfrac{9}{8!} + \cdots$    $= e$

8. $\dfrac{1}{1!} + \dfrac{2}{3!} + \dfrac{3}{5!} + \dfrac{4}{7!} + \dfrac{5}{9!} + \cdots$    $= \dfrac{1}{2}e$

9. $\dfrac{2}{3!} + \dfrac{4}{5!} + \dfrac{6}{7!} + \dfrac{8}{9!} + \dfrac{10}{11!} + \cdots$    $= e^{-1}$

10. $\dfrac{1}{2!} + \dfrac{2}{3!} + \dfrac{3}{4!} + \dfrac{4}{5!} + \dfrac{5}{6!} + \cdots$    $= 1$

11. $1 + \dfrac{2}{1!} + \dfrac{3}{2!} + \dfrac{4}{3!} + \dfrac{5}{4!} + \cdots$ $\qquad = 2e$

12. $\dfrac{1^2}{1!} + \dfrac{2^2}{2!} + \dfrac{3^2}{3!} + \dfrac{4^2}{4!} + \dfrac{5^2}{5!} + \cdots$ $\qquad = 2e$

13. $\dfrac{1^2}{2!} + \dfrac{2^2}{3!} + \dfrac{3^2}{4!} + \dfrac{4^2}{5!} + \dfrac{5^2}{6!} + \cdots$ $\qquad = (e-1)$

14. $1 + \dfrac{2^2}{1!} + \dfrac{3^2}{2!} + \dfrac{4^2}{3!} + \dfrac{5^2}{4!} + \cdots$ $\qquad = 5e$

15. $\dfrac{1 \cdot 3}{2!} + \dfrac{2 \cdot 4}{3!} + \dfrac{3 \cdot 5}{4!} + \dfrac{4 \cdot 6}{5!} + \cdots$ $\qquad = (e+1)$

16. $\dfrac{1 \cdot 3}{2!} + \dfrac{3 \cdot 5}{4!} + \dfrac{5 \cdot 7}{6!} + \dfrac{7 \cdot 9}{8!} + \cdots$ $\qquad = \dfrac{e^2 + 2e - 1}{2e}$

17. $\dfrac{2}{1!3} - \dfrac{3}{2!4} + \dfrac{4}{3!5} - \dfrac{5}{4!6} + \cdots$ $\qquad = 3\left( \dfrac{1}{2} - \dfrac{1}{e} \right)$

18. $\dfrac{1}{1!} + \dfrac{1+2}{2!} + \dfrac{1+2+3}{3!} + \dfrac{1+2+3+4}{4!} + \cdots$ $\; = \dfrac{3e}{2}$

19. $\dfrac{1(2^2+1)}{2!} + \dfrac{2(3^2+1)}{3!} + \dfrac{3(4^2+1)}{4!} + \cdots$ $\qquad = (3e+1)$

20. $\dfrac{1(2^2)}{2!} + \dfrac{2(3^2)}{3!} + \dfrac{3(4^2)}{4!} + \cdots$ $\qquad = (3e)$

21. $\dfrac{1(2^2-1)}{2!} + \dfrac{2(3^2-1)}{3!} + \dfrac{3(4^2-1)}{4!} + \cdots$ $\qquad = (3e-1)$

22. $\dfrac{e^{\frac{1}{1}} \cdot e^{\frac{1}{3}} \cdot e^{\frac{1}{5}} \cdot e^{\frac{1}{7}} \cdot e^{\frac{1}{9}} \cdot e^{\frac{1}{11}} \cdots}{e^{\frac{1}{2}} \cdot e^{\frac{1}{4}} \cdot e^{\frac{1}{6}} \cdot e^{\frac{1}{8}} \cdot e^{\frac{1}{10}} \cdot e^{\frac{1}{12}}}$ $\qquad = 2$

23. $e^{i\pi}$ $\qquad = -1$

24. $e^{i\pi} + 1$ $\qquad = 0$

# e-Series in $\Sigma$ notation

$$\sum_{1}^{\infty} \frac{1}{n!} = (e-1)$$

$$\sum_{1}^{\infty} \frac{n}{n!} = e$$

$$\sum_{1}^{\infty} \frac{n+1}{n!} = (2e-1)$$

$$\sum_{1}^{\infty} \frac{n^2}{n!} = 2e$$

$$\sum_{1}^{\infty} \frac{(n+1)^2}{n!} = (5e-1)$$

$$\sum_{1}^{\infty} \frac{1}{(n-1)!} = e$$

$$\sum_{1}^{\infty} \frac{n}{(n-1)!} = 2e$$

$$\sum_{1}^{\infty} \frac{n^2}{(n-1)!} = 5e$$

$$\sum_{1}^{\infty} \frac{1}{(n+1)!} = (e-2)$$

$$\sum_{1}^{\infty} \frac{n}{(n+1)!} = e^{0} = 1$$

$$\sum_{1}^{\infty} \frac{n^2}{(n+1)!} = (e-1)$$

$$\sum_{1}^{\infty} \frac{2n+1}{(2n)!} = (e-1)$$

$$\sum_{1}^{\infty} \frac{2n}{(2n+1)!} = e^{-1} = \frac{1}{e}$$

$$\sum_{0}^{\infty} \frac{(n+1)}{(2n+1)!} = \frac{1}{2}e$$

# Other Interesting Number Series

O, were the sum of these ...
countless and infinite ...

**William Shakespeare** (1564–1616)

§

Our minds are finite
Yet we are surrounded by possibilities
that are infinite
and the purpose of human life
is to grasp as much as we can
of that infinitude

**Alfred Whitehead** (1861–1947)

§

$$\frac{1}{2^1} + \frac{1}{2^2} + \frac{1}{2^3} + \frac{1}{2^4} + \frac{1}{2^5} + \frac{1}{2^6} + \cdots$$

$$= 1$$

$$\frac{9}{10^1} + \frac{9}{10^2} + \frac{9}{10^3} + \frac{9}{10^4} + \frac{9}{10^5} + \frac{9}{10^6} + \cdots$$

$$= 1$$

See Proof 12 → (page 156)

A BEAUTIFUL FORMULA for the simple integer 1. This series was introduced earlier on as our first convergent series.

How about the second series with 9's and 10's? Believe it or not, it also sums to 1. My granddaughter, Rebecca, looking at the 9's and 10's, refused to believe that it summed to 1, until I proved it to her with "the visual proof".

Do you see any common pattern in the two series? Can you come up with your own infinite series that sum to 1?

$$\frac{1}{2^1} + \frac{1}{2^2} + \frac{1}{2^3} + \cdots = 1$$

$$\frac{2}{3^1} + \frac{2}{3^2} + \frac{2}{3^3} + \cdots = 1$$

$$\frac{3}{4^1} + \frac{3}{4^2} + \frac{3}{4^3} + \cdots = 1$$

$$\frac{4}{5^1} + \frac{4}{5^2} + \frac{4}{5^3} + \cdots = 1$$

$$\frac{5}{6^1} + \frac{5}{6^2} + \frac{5}{6^3} + \cdots = 1$$

$$\frac{6}{7^1} + \frac{6}{7^2} + \frac{6}{7^3} + \cdots = 1$$

$$\frac{7}{8^1} + \frac{7}{8^2} + \frac{7}{8^3} + \cdots = 1$$

$$\frac{8}{9^1} + \frac{8}{9^2} + \frac{8}{9^3} + \cdots = 1$$

$$\frac{9}{10^1} + \frac{9}{10^2} + \frac{9}{10^3} + \cdots = 1$$

See Proof 12 → (page 156)

ISN'T IT AMAZING how many different infinite series there are for representing 1. In fact if you think about it, there is an infinite (yes, literally infinite) number of infinite series for representing 1.

Can you work out the general formula?

$$\frac{3}{2} - \frac{3}{2^2} + \frac{3}{2^3} - \frac{3}{2^4} + \cdots = 1$$

$$\frac{4}{3} - \frac{4}{3^2} + \frac{4}{3^3} - \frac{4}{3^4} + \cdots = 1$$

$$\frac{5}{4} - \frac{5}{4^2} + \frac{5}{4^3} - \frac{5}{4^4} + \cdots = 1$$

$$\frac{6}{5} - \frac{6}{5^2} + \frac{6}{5^3} - \frac{6}{5^4} + \cdots = 1$$

$$\frac{7}{6} - \frac{7}{6^2} + \frac{7}{6^3} - \frac{7}{6^4} + \cdots = 1$$

$$\frac{8}{7} - \frac{8}{7^2} + \frac{8}{7^3} - \frac{8}{7^4} + \cdots = 1$$

$$\frac{9}{8} - \frac{9}{8^2} + \frac{9}{8^3} - \frac{9}{8^4} + \cdots = 1$$

$$\frac{10}{9} - \frac{10}{9^2} + \frac{10}{9^3} - \frac{10}{9^4} + \cdots = 1$$

See Proof 12 → (page 156)

How ABOUT ANOTHER infinite class of infinite series that sums to 1? Can you work out the general formula for this one too?

Note especially the alternating terms (positive followed by negative) in the series. Note also that the class does not begin with the $(\frac{2}{1})$ series. Do you know why?

$$\frac{1}{2!} + \frac{2}{3!} + \frac{3}{4!} + \frac{4}{5!} + \frac{5}{6!} + \frac{6}{7!} + \cdots$$

$$= 1$$

See Proof 28 → (page 190)

REMEMBER SEEING THIS beautiful series before? You saw it among the e infinite series. This is one of the more mind-boggling of equations in this book — all factorials and no e.

Isn't it a beautiful way to represent 1 as an infinite series!

The series is re-introduced here to get you acquainted with seeing the progressively increasing integers in the numerators because we are going to have lots of fun with them later in this chapter.

$$\frac{1}{1\cdot 2} + \frac{1}{2\cdot 3} + \frac{1}{3\cdot 4} + \frac{1}{4\cdot 5} + \frac{1}{5\cdot 6} + \cdots$$

$$= 1$$

See Proof 1 → (page 145)

YET ANOTHER SIMPLE infinite series for 1. How can something so simple be equal to the sum of so many different but equally beautiful and simple infinite series? The proof is so simple that Rebecca could follow and understand it.

Indeed it is so simple that it has been given the place of honour as "Proof 1" — the easiest proof in this book.

Can you do it without looking at the answer at the back of the book?

$$\frac{1}{1 \cdot 2} + \frac{1}{3 \cdot 4} + \frac{1}{5 \cdot 6} + \frac{1}{7 \cdot 8} + \frac{1}{9 \cdot 10} + \cdots$$

$$= \log_{\text{natural}} 2$$

See Proof 3 → (page 147)

CHANGING THE PREVIOUS equation a little, we get this beautiful equation of rational reciprocals of integers, summing to an irrational constant, the natural logarithm of 2. We saw a slightly different infinite series with the same sum earlier on. A simple proof will demonstrate that these two different infinite series are really different facets of the same equation.

$$\frac{1}{1 \cdot 3} + \frac{1}{3 \cdot 5} + \frac{1}{5 \cdot 7} + \frac{1}{7 \cdot 9} + \frac{1}{9 \cdot 11} + \cdots$$

$$= \frac{1}{2}$$

$$\frac{1}{1 \cdot 5} + \frac{1}{3 \cdot 7} + \frac{1}{5 \cdot 9} + \frac{1}{7 \cdot 11} + \frac{1}{9 \cdot 13} + \cdots$$

$$= \frac{1}{3}$$

See Proofs 5, 7 → (pages 149, 151)

Two MORE BEAUTIFUL sisters for the previous series. More petite now, summing to smaller rational fractions, but no less beautiful.

(Can you create your own infinite series for other simple fractions?)

$$\frac{1}{1\cdot2\cdot3} + \frac{1}{2\cdot3\cdot4} + \frac{1}{3\cdot4\cdot5} + \frac{1}{4\cdot5\cdot6} + \cdots$$

$$= \frac{1}{4}$$

$$\frac{1}{1\cdot3\cdot5} + \frac{1}{3\cdot5\cdot7} + \frac{1}{5\cdot7\cdot9} + \frac{1}{7\cdot9\cdot11} + \cdots$$

$$= \frac{1}{12}$$

See Proofs 8, 9 → (pages 152, 153)

COUSINS OF THE previous equations.

$$\frac{1}{1\cdot 2\cdot 3}+\frac{2}{2\cdot 3\cdot 4}+\frac{3}{3\cdot 4\cdot 5}+\frac{4}{4\cdot 5\cdot 6}+\cdots$$

$$=\frac{1}{2}$$

$$\frac{1}{1\cdot 3\cdot 5}+\frac{2}{3\cdot 5\cdot 7}+\frac{3}{5\cdot 7\cdot 9}+\frac{4}{7\cdot 9\cdot 11}+\cdots$$

$$=\frac{1}{8}$$

See Proofs 10, 11 → (pages 154, 155)

DISTANT COUSINS.

Note especially the progressively increasing integers in the numerators. We'll see them again in due course.

$$\frac{1}{1\cdot 2}+\frac{1}{2\cdot 3}+\frac{1}{3\cdot 4}+\cdots \qquad\qquad = 1$$

$$\frac{1}{1\cdot 3}+\frac{1}{3\cdot 5}+\frac{1}{5\cdot 7}+\cdots \qquad\qquad = \frac{1}{2}$$

$$\frac{1}{1\cdot 4}+\frac{1}{4\cdot 7}+\frac{1}{7\cdot 10}+\cdots \qquad\qquad = \frac{1}{3}$$

$$\cdots\cdots\cdots\cdots\cdots\cdots\cdots\cdots\cdots\cdots\cdots\cdots$$

$$\cdots\cdots\cdots\cdots\cdots\cdots\cdots\cdots\cdots\cdots\cdots\cdots$$

$$\frac{1}{1\cdot n}+\frac{1}{n(2n-1)}+\frac{1}{(2n-1)(3n-2)}+\cdots = \frac{1}{(n-1)}$$

See Proof 6 → (page 150)

Here is another general equation for an infinite class of infinite series, giving simple rational fractional sums.

Amazing isn't it, that we can have an infinite series that sums to as small a fraction as we like, simply by using the largest $n$ that we like.

$$\frac{1}{1\cdot 2}+\frac{1}{2\cdot 3}+\frac{1}{3\cdot 4}+\frac{1}{4\cdot 5}+\frac{1}{5\cdot 6}+\cdots \qquad =1$$

$$\frac{1}{2\cdot 3}+\frac{1}{3\cdot 4}+\frac{1}{4\cdot 5}+\frac{1}{5\cdot 6}+\cdots \qquad =\frac{1}{2}$$

$$\frac{1}{3\cdot 4}+\frac{1}{4\cdot 5}+\frac{1}{5\cdot 6}+\cdots \qquad =\frac{1}{3}$$

....................................................................

....................................................................

$$\frac{1}{n(n+1)}+\frac{1}{(n+1)(n+2)}+\frac{1}{(n+2)(n+3)}+\cdots =\frac{1}{n}$$

See Proof 2 → (page 146)

ANOTHER GENERAL EQUATION for another infinite class of infinite series with simple rational fractions for sums.

Can you see the patterns and work out the sums mentally after the first series? Can you create your own infinite series for a simple fraction, say $\frac{1}{1,000,000}$?

$$\frac{1}{1} - \frac{1}{2} + \frac{1}{4} - \frac{1}{8} + \frac{1}{16} - \frac{1}{32} + \cdots = \frac{2}{3} \quad = 0.6666\ldots$$

$$\frac{1}{1} + \frac{1}{2} + \frac{1}{4} + \frac{1}{8} + \frac{1}{16} + \frac{1}{32} + \cdots = 2$$

$$\frac{1}{1} - \frac{1}{2} + \frac{1}{3} - \frac{1}{4} + \frac{1}{5} - \frac{1}{6} + \cdots \quad = \ln 2 \quad = 0.6931\ldots$$

$$\frac{1}{1} + \frac{1}{2} + \frac{1}{3} + \frac{1}{4} + \frac{1}{5} + \frac{1}{6} + \cdots \quad \rightarrow \infty$$

$$\frac{1}{1} - \frac{1}{3} + \frac{1}{5} - \frac{1}{7} + \frac{1}{9} - \frac{1}{11} + \cdots \quad = \frac{\pi}{4} \quad = 0.7854\ldots$$

$$\frac{1}{1} + \frac{1}{3} + \frac{1}{5} + \frac{1}{7} + \frac{1}{9} + \frac{1}{11} + \cdots \quad \rightarrow \infty$$

See Proof 12 → (page 156)

THE SUM OF the Geometric Series with alternating positive and negative terms is $\frac{2}{3}$ (0.6666 ...). The famous Geometric Series with all positive terms sums to 2, three times that of its sister series with alternating terms.

The sum of the Logarithm Series (with alternating terms) is ln 2 (0.6931 ...), very similar to the sum for the series above summing to 0.6666 .... But its sister series with all positive terms, now becomes the famous Harmonic Series, with its sum tending to infinity.

Similarly, the sum of the beautiful Liebniz-Gregory series (with alternating terms) is $\frac{\pi}{4}$ (0.7854 ...). Again its sister series with all positive terms is transformed into the "Half" Harmonic series with reciprocals of odd integers. Its sum tends to infinity.

Isn't it amazing in the field of infinite series how a small difference sometimes makes for a difference of infinity. Hence one has to be very careful when dealing with infinity and infinite sums. Intuition, or common sense, is often a poor and misleading guide.

$$\frac{1}{2^1} + \frac{1}{2^2} + \frac{1}{2^3} + \frac{1}{2^4} + \cdots \ = 1$$

$$\frac{1}{2^1} + \frac{2}{2^2} + \frac{3}{2^3} + \frac{4}{2^4} + \cdots \ = 2$$

$$1 + \frac{1}{1!} + \frac{1}{2!} + \frac{1}{3!} + \frac{1}{4!} + \cdots = e$$

$$1 + \frac{2}{1!} + \frac{3}{2!} + \frac{4}{3!} + \frac{5}{4!} + \cdots = 2e$$

See Proofs 13, 29 $\rightarrow$ (pages 164, 192)

SEE THE SYMMETRY in the Geometric Series and its sister series with progressively increasing integers in the numerators. Isn't it beautiful? And an extremely interesting but still simple proof.

The corresponding sister series for the e series sums to 2e, also twice that of the e-series.

Which makes one wonder if there is such a pattern for other series, to sum to twice the original sum if we change their numerators from 1 to progressively increasing integers, 1, 2, 3, 4 ... *n*?

$$\frac{1}{1\cdot2}+\frac{1}{2\cdot3}+\frac{1}{3\cdot4}+\frac{1}{4\cdot5}+\cdots \quad = 1$$

$$\frac{1}{1\cdot2}+\frac{2}{2\cdot3}+\frac{3}{3\cdot4}+\frac{4}{4\cdot5}+\cdots \quad \to \infty$$

$$\frac{1}{1\cdot2}+\frac{1}{3\cdot4}+\frac{1}{5\cdot6}+\frac{1}{7\cdot8}+\cdots \quad = \ln 2\,(0.69\ldots)$$

$$\frac{1}{1\cdot2}+\frac{2}{3\cdot4}+\frac{3}{5\cdot6}+\frac{4}{7\cdot8}+\cdots \quad \to \infty$$

$$\frac{1}{1\cdot3}+\frac{1}{3\cdot5}+\frac{1}{5\cdot7}+\frac{1}{7\cdot9}+\cdots \quad = \frac{1}{2}\,(0.50\ldots)$$

$$\frac{1}{1\cdot3}+\frac{2}{3\cdot5}+\frac{3}{5\cdot7}+\frac{4}{7\cdot9}+\cdots \quad \to \infty$$

$$\frac{1}{1\cdot3}+\frac{1}{5\cdot7}+\frac{1}{9\cdot11}+\frac{1}{13\cdot15}+\cdots = \frac{\pi}{2\cdot4}\,(0.39\ldots)$$

$$\frac{1}{1\cdot3}+\frac{2}{5\cdot7}+\frac{3}{9\cdot11}+\frac{4}{13\cdot15}+\cdots \to \infty$$

See Proofs 16–19 → (pages 168–171)

LOOKING AT THE four pairs of infinite series on the opposite page, it is clear that our working hypothesis formed from the previous equations is not valid. Four beautiful series summing to 1, ln 2, $\frac{1}{2}$, and $\frac{\pi}{2 \cdot 4}$, undergo extreme transformation when their numerators of 1 are changed to 1, 2, 3, 4 ... $n$ progressively. Not only do they not sum to double the sum of their sister series, they do not sum to any constant at all, not even to a large or super large one. They are no longer convergent. They are now divergent, and their sums individually tend to infinity! WOW! They say "a miss is as good as a mile". In mathematics, "a minor change can be as good as a difference of infinity".

$$\frac{1}{1\cdot 2}+\frac{1}{2\cdot 3}+\frac{1}{3\cdot 4}+\frac{1}{4\cdot 5}+\frac{1}{5\cdot 6}+\cdots \qquad\qquad =1$$

$$\frac{1}{1\cdot 2}+\frac{2}{2\cdot 3}+\frac{3}{3\cdot 4}+\frac{4}{4\cdot 5}+\frac{5}{5\cdot 6}+\cdots \qquad\qquad \to \infty$$

$$\frac{1}{2\cdot 3}+\frac{2}{3\cdot 4}+\frac{3}{4\cdot 5}+\frac{4}{5\cdot 6}+\cdots \qquad\qquad \to \infty$$

$$\frac{1}{3\cdot 4}+\frac{2}{4\cdot 5}+\frac{3}{5\cdot 6}+\cdots \qquad\qquad \to \infty$$

.......................................................

.......................................................

$$\frac{1}{1,000,000\cdot 1,000,001}+\frac{2}{1,000,001\cdot 1,000,002}+\cdots \to \infty$$

$$\frac{1}{n\cdot (n+1)}+\frac{2}{(n+1)\cdot (n+2)}+\cdots \qquad\qquad \to \infty$$

See Proof 20 → (page 173)

WE SAW HOW a minor change in an infinite series can make a difference of infinity.

Let us now show an example of the other extreme — where the sum of an infinite series tends to infinity, regardless of how much we "chop and change", even to the extent of chopping off a million terms, and moving the numerators 1, 2, 3 ... down the rest of the subsequent terms of the series.

$$\frac{1}{1\cdot 2} + \frac{1}{2\cdot 3} + \frac{1}{3\cdot 4} + \frac{1}{4\cdot 5} + \cdots \qquad = 1$$

$$\frac{1}{1\cdot 2} + \frac{2}{2\cdot 3} + \frac{3}{3\cdot 4} + \frac{4}{4\cdot 5} + \cdots \qquad \rightarrow \infty$$

$$\frac{1}{1\cdot 3} + \frac{2}{3\cdot 5} + \frac{3}{5\cdot 7} + \frac{4}{7\cdot 9} + \cdots \qquad \rightarrow \infty$$

$$\frac{1}{1\cdot 4} + \frac{2}{4\cdot 7} + \frac{3}{7\cdot 10} + \frac{4}{10\cdot 13} + \cdots \qquad \rightarrow \infty$$

$$\cdots\cdots\cdots\cdots\cdots\cdots\cdots\cdots\cdots\cdots\cdots\cdots\cdots$$

$$\cdots\cdots\cdots\cdots\cdots\cdots\cdots\cdots\cdots\cdots\cdots\cdots\cdots$$

$$\frac{1}{1\cdot n} + \frac{2}{n(2n-1)} + \frac{3}{(2n-1)(3n-2)} + \cdots \rightarrow \infty$$

See Proof 21 → (page 174)

LET US BRING this section on infinite series to a close with a big bang of infinities — indeed an infinity of infinite series, the sums of which all tend to infinity!

No matter how small the individual terms may be, certain series always end up with sums tending to infinity, as illustrated by the class of infinite series opposite. Even when $n$ is very large, and hence the terms beginning $\frac{1}{1 \cdot n}$ are all very small, their sums still tend to infinity. Isn't that amazing!

This final class of infinite series also serves to introduce us to the work of the 19th century German mathematics genius, Georg Cantor (1845–1918). Cantor was the first to open wide the door into the wild, wacky, wondrous world of infinities. (Yes, there are many classes of infinities, different one from another. Indeed there is an infinite hierarchy of infinities, with some infinities greater than others!) Cantor was also the first to prove that there are as many odd integers as there are ordinary integers (not half, as one would intuitively expect), and that there are as many rational fractions between 0 and 1, as there are ordinary integers!

Isn't infinity amazing! Isn't it mysterious! Isn't this a beautiful way to end the section on the infinite series of $\pi$, e and other interesting numbers!

# Other Interesting Number Series

1. $\dfrac{1}{2^1} + \dfrac{1}{2^2} + \dfrac{1}{2^3} + \dfrac{1}{2^4} + \dfrac{1}{2^5} + \cdots$ $\qquad = 1$

2. $\dfrac{2}{3^1} + \dfrac{2}{3^2} + \dfrac{2}{3^3} + \dfrac{2}{3^4} + \dfrac{2}{3^5} + \cdots$ $\qquad = 1$

.................................................................

3. $\dfrac{9}{10^1} + \dfrac{9}{10^2} + \dfrac{9}{10^3} + \dfrac{9}{10^4} + \dfrac{9}{10^5} + \cdots$ $\qquad = 1$

4. $\dfrac{n}{(n+1)^1} + \dfrac{n}{(n+1)^2} + \dfrac{n}{(n+1)^3} + \dfrac{n}{(n+1)^4} + \cdots$ $\qquad = 1$

5. $\dfrac{3}{2^1} - \dfrac{3}{2^2} + \dfrac{3}{2^3} - \dfrac{3}{2^4} + \dfrac{3}{2^5} - \cdots$ $\qquad = 1$

6. $\dfrac{4}{3^1} - \dfrac{4}{3^2} + \dfrac{4}{3^3} - \dfrac{4}{3^4} + \dfrac{4}{3^5} - \cdots$ $\qquad = 1$

.................................................................

7. $\dfrac{10}{9^1} - \dfrac{10}{9^2} + \dfrac{10}{9^3} - \dfrac{10}{9^4} + \dfrac{10}{9^5} - \cdots$ $\qquad = 1$

8. $\dfrac{n+1}{n^1} - \dfrac{n+1}{n^2} + \dfrac{n+1}{n^3} - \dfrac{n+1}{n^4} + \cdots$ $\qquad = 1$

9. $\dfrac{1}{2!} + \dfrac{2}{3!} + \dfrac{3}{4!} + \dfrac{4}{5!} + \dfrac{5}{6!} + \cdots$ $\qquad = 1$

10. $\dfrac{1}{1 \cdot 2} + \dfrac{1}{2 \cdot 3} + \dfrac{1}{3 \cdot 4} + \dfrac{1}{4 \cdot 5} + \cdots$ $\qquad = 1$

11. $\dfrac{1}{1\cdot 2}+\dfrac{1}{3\cdot 4}+\dfrac{1}{5\cdot 6}+\dfrac{1}{7\cdot 8}+\cdots \qquad = \ln 2$

12. $\dfrac{1}{1\cdot 3}+\dfrac{1}{3\cdot 5}+\dfrac{1}{5\cdot 7}+\dfrac{1}{7\cdot 9}+\cdots \qquad = \dfrac{1}{2}$

13. $\dfrac{1}{1\cdot 3}+\dfrac{1}{5\cdot 7}+\dfrac{1}{9\cdot 11}+\dfrac{1}{13\cdot 15}+\cdots \qquad = \dfrac{\pi}{2\cdot 4}$

14. $\dfrac{1}{1\cdot 5}+\dfrac{1}{3\cdot 7}+\dfrac{1}{5\cdot 9}+\dfrac{1}{7\cdot 11}+\cdots \qquad = \dfrac{1}{3}$

15. $\dfrac{1}{1\cdot 2\cdot 3}+\dfrac{1}{2\cdot 3\cdot 4}+\dfrac{1}{3\cdot 4\cdot 5}+\cdots \qquad = \dfrac{1}{4}$

16. $\dfrac{1}{1\cdot 3\cdot 5}+\dfrac{1}{3\cdot 5\cdot 7}+\dfrac{1}{5\cdot 7\cdot 9}+\cdots \qquad = \dfrac{1}{12}$

17. $\dfrac{1}{1\cdot 2\cdot 3}+\dfrac{2}{2\cdot 3\cdot 4}+\dfrac{3}{3\cdot 4\cdot 5}+\cdots \qquad = \dfrac{1}{2}$

18. $\dfrac{1}{1\cdot 3\cdot 5}+\dfrac{2}{3\cdot 5\cdot 7}+\dfrac{3}{5\cdot 7\cdot 9}+\cdots \qquad = \dfrac{1}{8}$

19. $\dfrac{1}{1\cdot 2}+\dfrac{1}{2\cdot 3}+\dfrac{1}{3\cdot 4}+\dfrac{1}{4\cdot 5}+\cdots \qquad = 1$

20. $\dfrac{1}{1\cdot 3}+\dfrac{1}{3\cdot 5}+\dfrac{1}{5\cdot 7}+\dfrac{1}{7\cdot 9}+\cdots \qquad = \dfrac{1}{2}$

...................................................

21. $\dfrac{1}{1\cdot n}+\dfrac{1}{n(2n-1)}+\dfrac{1}{(2n-1)(3n-2)}+\cdots \qquad = \dfrac{1}{(n-1)}$

22. $\dfrac{1}{1\cdot 2}+\dfrac{1}{2\cdot 3}+\dfrac{1}{3\cdot 4}+\dfrac{1}{4\cdot 5}+\cdots \qquad = 1$

23. $\dfrac{1}{2\cdot 3}+\dfrac{1}{3\cdot 4}+\dfrac{1}{4\cdot 5}+\cdots \qquad = \dfrac{1}{2}$

...................................................

24. $\dfrac{1}{n(n+1)} + \dfrac{1}{(n+1)(n+2)} + \dfrac{1}{(n+2)(n+3)} + \cdots \qquad = \dfrac{1}{n}$

25. $\dfrac{1}{1} - \dfrac{1}{2} + \dfrac{1}{4} - \dfrac{1}{8} + \dfrac{1}{16} - \dfrac{1}{32} + \cdots \qquad = \dfrac{2}{3}\,(0.6666\ldots)$

26. $\dfrac{1}{1} + \dfrac{1}{2} + \dfrac{1}{4} + \dfrac{1}{8} + \dfrac{1}{16} + \dfrac{1}{32} + \cdots \qquad = 2$

27. $\dfrac{1}{1} - \dfrac{1}{2} + \dfrac{1}{3} - \dfrac{1}{4} + \dfrac{1}{5} - \dfrac{1}{6} + \cdots \qquad = \ln 2\,(0.6931\ldots)$

28. $\dfrac{1}{1} + \dfrac{1}{2} + \dfrac{1}{3} + \dfrac{1}{4} + \dfrac{1}{5} + \dfrac{1}{6} + \cdots \qquad \to \infty$

29. $\dfrac{1}{1} - \dfrac{1}{3} + \dfrac{1}{5} - \dfrac{1}{7} + \dfrac{1}{9} - \dfrac{1}{11} + \cdots \qquad = \dfrac{\pi}{4}\,(0.7854\ldots)$

30. $\dfrac{1}{1} + \dfrac{1}{3} + \dfrac{1}{5} + \dfrac{1}{7} + \dfrac{1}{9} + \dfrac{1}{11} + \cdots \qquad \to \infty$

31. $\dfrac{1}{2^1} + \dfrac{1}{2^2} + \dfrac{1}{2^3} + \dfrac{1}{2^4} + \cdots \qquad = 1$

32. $\dfrac{1}{2^1} + \dfrac{2}{2^2} + \dfrac{3}{2^3} + \dfrac{4}{2^4} + \cdots \qquad = 2$

33. $1 + \dfrac{1}{1!} + \dfrac{1}{2!} + \dfrac{1}{3!} + \dfrac{1}{4!} + \cdots \qquad = e$

34. $1 + \dfrac{2}{1!} + \dfrac{3}{2!} + \dfrac{4}{3!} + \dfrac{5}{4!} + \cdots \qquad = 2e$

35. $\dfrac{1}{1 \cdot 2} + \dfrac{1}{2 \cdot 3} + \dfrac{1}{3 \cdot 4} + \dfrac{1}{4 \cdot 5} + \cdots \qquad = 1$

36. $\dfrac{1}{1 \cdot 2} + \dfrac{2}{2 \cdot 3} + \dfrac{3}{3 \cdot 4} + \dfrac{4}{4 \cdot 5} + \cdots \qquad \to \infty$

37. $\dfrac{1}{1 \cdot 2} + \dfrac{1}{3 \cdot 4} + \dfrac{1}{5 \cdot 6} + \dfrac{1}{7 \cdot 8} + \cdots \qquad = \ln 2\,(0.6931\ldots)$

38. $\dfrac{1}{1\cdot 2}+\dfrac{2}{3\cdot 4}+\dfrac{3}{5\cdot 6}+\dfrac{4}{7\cdot 8}+\cdots$ $\qquad\rightarrow\infty$

39. $\dfrac{1}{1\cdot 3}+\dfrac{1}{3\cdot 5}+\dfrac{1}{5\cdot 7}+\dfrac{1}{7\cdot 9}+\cdots$ $\qquad =\dfrac{1}{2}\,(0.5000\ldots)$

40. $\dfrac{1}{1\cdot 3}+\dfrac{2}{3\cdot 5}+\dfrac{3}{5\cdot 7}+\dfrac{4}{7\cdot 9}+\cdots$ $\qquad\rightarrow\infty$

41. $\dfrac{1}{1\cdot 3}+\dfrac{1}{5\cdot 7}+\dfrac{1}{9\cdot 11}+\dfrac{1}{13\cdot 15}+\cdots$ $\qquad =\dfrac{\pi}{2\cdot 4}\,(0.3926\ldots)$

42. $\dfrac{1}{1\cdot 3}+\dfrac{2}{5\cdot 7}+\dfrac{3}{9\cdot 11}+\dfrac{4}{13\cdot 15}+\cdots$ $\qquad\rightarrow\infty$

43. $\dfrac{1}{1\cdot 2}+\dfrac{1}{2\cdot 3}+\dfrac{1}{3\cdot 4}+\dfrac{1}{4\cdot 5}+\cdots$ $\qquad =1$

44. $\dfrac{1}{1\cdot 2}+\dfrac{2}{2\cdot 3}+\dfrac{3}{3\cdot 4}+\dfrac{4}{4\cdot 5}+\cdots$ $\qquad\rightarrow\infty$

45. $\dfrac{1}{2\cdot 3}+\dfrac{2}{3\cdot 4}+\dfrac{3}{4\cdot 5}+\cdots$ $\qquad\rightarrow\infty$

46. $\dfrac{1}{3\cdot 4}+\dfrac{2}{4\cdot 5}+\cdots$ $\qquad\rightarrow\infty$

$\cdots\cdots\cdots\cdots\cdots\cdots\cdots\cdots\cdots\cdots\cdots\cdots\cdots\cdots\cdots\cdots\cdots$

47. $\dfrac{1}{1,000,000\cdot 1,000,001}+\dfrac{2}{1,000,001\cdot 1,000,002}+\cdots\rightarrow\infty$

48. $\dfrac{1}{n\cdot (n+1)}+\dfrac{2}{(n+1)\cdot (n+2)}+\cdots$ $\qquad\rightarrow\infty$

49. $\dfrac{1}{1\cdot 2}+\dfrac{1}{2\cdot 3}+\dfrac{1}{3\cdot 4}+\dfrac{1}{4\cdot 5}+\cdots$ $\qquad =1$

50. $\dfrac{1}{1\cdot 2}+\dfrac{2}{2\cdot 3}+\dfrac{3}{3\cdot 4}+\dfrac{4}{4\cdot 5}+\cdots$ $\qquad\rightarrow\infty$

51. $\dfrac{1}{1\cdot 3}+\dfrac{2}{3\cdot 5}+\dfrac{3}{5\cdot 7}+\dfrac{4}{7\cdot 9}+\cdots$ $\qquad\to\infty$

52. $\dfrac{1}{1\cdot 4}+\dfrac{2}{4\cdot 7}+\dfrac{3}{7\cdot 10}+\dfrac{4}{10\cdot 13}+\cdots$ $\qquad\to\infty$

......................................................

53. $\dfrac{1}{1\cdot n}+\dfrac{2}{n(2n-1)}+\dfrac{3}{(2n+1)(3n-2)}+\cdots$ $\qquad\to\infty$

*Easy Proofs*

Put not yourself into amazement
how these things should be
all difficulties are but easy
when they are known

**William Shakespeare** (1564–1616)

§

A thing is obvious mathematically
after you see it

**R.D. Carmichael** (1879–1967)

§

The most incomprehensible thing
about the world
is that it is comprehensible

**Albert Einstein** (1879–1955)

§

# Proof 1

$$\frac{1}{1\cdot 2}+\frac{1}{2\cdot 3}+\frac{1}{3\cdot 4}+\frac{1}{4\cdot 5}+\cdots=1$$

$$\frac{1}{1\cdot 2}=\left(\frac{1}{1}-\frac{1}{2}\right)$$

$$\frac{1}{2\cdot 3}=\left(\frac{1}{2}-\frac{1}{3}\right)$$

$$\frac{1}{3\cdot 4}=\left(\frac{1}{3}-\frac{1}{4}\right)$$

..................

..................

$$\frac{1}{1\cdot 2}+\frac{1}{2\cdot 3}+\frac{1}{3\cdot 4}+\cdots$$

$$=\left(\frac{1}{1}-\frac{1}{2}\right)+\left(\frac{1}{2}-\frac{1}{3}\right)+\left(\frac{1}{3}-\frac{1}{4}\right)+\cdots$$

$$=1$$

# Proof 2

$$\frac{1}{1\cdot 2}+\frac{1}{2\cdot 3}+\frac{1}{3\cdot 4}+\frac{1}{4\cdot 5}+\cdots =1$$

$$\frac{1}{2\cdot 3}+\frac{1}{3\cdot 4}+\frac{1}{4\cdot 5}+\cdots =1-\frac{1}{2}\qquad =\frac{1}{2}$$

$$\frac{1}{3\cdot 4}+\frac{1}{4\cdot 5}+\cdots =1-\frac{1}{2}-\frac{1}{6}=\frac{1}{3}$$

$$\frac{1}{n(n+1)}+\frac{1}{(n+1)(n+2)}+\frac{1}{(n+2)(n+3)}+\cdots \qquad =\frac{1}{n}$$

$$\frac{1}{n(n+1)}+\frac{1}{(n+1)(n+2)}+\cdots$$

$$=\left(\frac{1}{n}-\frac{1}{n+1}\right)+\left(\frac{1}{n+1}-\frac{1}{n+2}\right)+\cdots$$

$$=\frac{1}{n}$$

# Proof 3

$$\frac{1}{1\cdot 2}+\frac{1}{3\cdot 4}+\frac{1}{5\cdot 6}+\frac{1}{7\cdot 8}+\cdots = \log_{\text{natural}} 2$$

$$\frac{1}{1\cdot 2}=\left(\frac{1}{1}-\frac{1}{2}\right)$$

$$\frac{1}{3\cdot 4}=\left(\frac{1}{3}-\frac{1}{4}\right)$$

$$\frac{1}{5\cdot 6}=\left(\frac{1}{5}-\frac{1}{6}\right)$$

$$\text{Sum}=\left(\frac{1}{1}-\frac{1}{2}\right)+\left(\frac{1}{3}-\frac{1}{4}\right)+\left(\frac{1}{5}-\frac{1}{6}\right)+\cdots$$

$$=1-\frac{1}{2}+\frac{1}{3}-\frac{1}{4}+\frac{1}{5}-\frac{1}{6}+\cdots$$

$$=\log_{\text{natural}} 2$$

# Proof 4

$$\frac{1}{1 \cdot 3} + \frac{1}{5 \cdot 7} + \frac{1}{9 \cdot 11} + \frac{1}{13 \cdot 15} + \cdots = \frac{\pi}{2 \cdot 4}$$

$$\frac{1}{1 \cdot 3} = \frac{1}{2}\left(\frac{1}{1} - \frac{1}{3}\right)$$

$$\frac{1}{5 \cdot 7} = \frac{1}{2}\left(\frac{1}{5} - \frac{1}{7}\right)$$

$$\frac{1}{9 \cdot 11} = \frac{1}{2}\left(\frac{1}{9} - \frac{1}{11}\right)$$

$$\text{Sum} = \frac{1}{2}\left(\frac{1}{1} - \frac{1}{3}\right) + \frac{1}{2}\left(\frac{1}{5} - \frac{1}{7}\right) + \frac{1}{2}\left(\frac{1}{9} - \frac{1}{11}\right) + \cdots$$

$$= \frac{1}{2}\left(1 - \frac{1}{3} + \frac{1}{5} - \frac{1}{7} + \frac{1}{9} + \cdots\right)$$

$$= \frac{1}{2}\left(\frac{\pi}{4}\right)$$

$$= \frac{\pi}{2 \cdot 4}$$

# Proof 5

$$\frac{1}{1\cdot 3}+\frac{1}{3\cdot 5}+\frac{1}{5\cdot 7}+\frac{1}{7\cdot 9}+\cdots=\frac{1}{2}$$

$$\frac{1}{1\cdot 3}=\frac{1}{2}\left(\frac{1}{1}-\frac{1}{3}\right)$$

$$\frac{1}{3\cdot 5}=\frac{1}{2}\left(\frac{1}{3}-\frac{1}{5}\right)$$

$$\frac{1}{5\cdot 7}=\frac{1}{2}\left(\frac{1}{5}-\frac{1}{7}\right)$$

$$\text{Sum}=\frac{1}{2}\left(\frac{1}{1}-\frac{1}{3}\right)+\frac{1}{2}\left(\frac{1}{3}-\frac{1}{5}\right)+\frac{1}{2}\left(\frac{1}{5}-\frac{1}{7}\right)+\cdots$$

$$=\frac{1}{2}(1)$$

$$=\frac{1}{2}$$

# Proof 6

$$\frac{1}{1\cdot 2}+\frac{1}{2\cdot 3}+\frac{1}{3\cdot 4}+\cdots \qquad\qquad =1$$

$$\frac{1}{1\cdot 3}+\frac{1}{3\cdot 5}+\frac{1}{5\cdot 7}+\cdots \qquad\qquad =\frac{1}{2}$$

$$\frac{1}{1\cdot 4}+\frac{1}{4\cdot 7}+\frac{1}{7\cdot 10}+\cdots \qquad\qquad =\frac{1}{3}$$

$$\cdots\cdots\cdots\cdots\cdots\cdots\cdots\cdots\cdots$$

$$\cdots\cdots\cdots\cdots\cdots\cdots\cdots\cdots\cdots$$

$$\frac{1}{1\cdot n}+\frac{1}{n(2n-1)}+\frac{1}{(2n-1)(3n-2)}+\cdots =\frac{1}{(n-1)}$$

$$\frac{1}{1\cdot n}=\frac{1}{(n-1)}\left(\frac{1}{1}-\frac{1}{n}\right)$$

$$\frac{1}{n\cdot(2n-1)}=\frac{1}{(n-1)}\left(\frac{1}{n}-\frac{1}{(2n-1)}\right)$$

$$\text{Sum}=\frac{1}{(n-1)}\left[\left(\frac{1}{1}-\frac{1}{n}\right)+\left(\frac{1}{n}-\frac{1}{(2n-1)}\right)+\cdots\right]$$

$$=\frac{1}{(n-1)}(1)$$

$$=\frac{1}{(n-1)}$$

# Proof 7

$$\frac{1}{1\cdot 5}+\frac{1}{3\cdot 7}+\frac{1}{5\cdot 9}+\frac{1}{7\cdot 11}+\cdots = \frac{1}{3}$$

$$\frac{1}{1\cdot 5}=\frac{1}{4}\left(\frac{1}{1}-\frac{1}{5}\right)$$

$$\frac{1}{3\cdot 7}=\frac{1}{4}\left(\frac{1}{3}-\frac{1}{7}\right)$$

$$\frac{1}{5\cdot 9}=\frac{1}{4}\left(\frac{1}{5}-\frac{1}{9}\right)$$

$$\text{Sum}=\frac{1}{4}\left[\left(\frac{1}{1}-\frac{1}{5}\right)+\left(\frac{1}{3}-\frac{1}{7}\right)+\left(\frac{1}{5}-\frac{1}{9}\right)+\cdots\right]$$

$$=\frac{1}{4}\left[\frac{1}{1}+\frac{1}{3}\right]$$

$$=\frac{1}{3}$$

Can you create your own $\frac{1}{x\cdot y}$ series and find their sums?

# Proof 8

$$\frac{1}{1\cdot 2\cdot 3}+\frac{1}{2\cdot 3\cdot 4}+\frac{1}{3\cdot 4\cdot 5}+\cdots=\frac{1}{4}$$

$$\frac{1}{1\cdot 2\cdot 3}=\frac{1}{2}\left(\frac{1}{1\cdot 2}-\frac{1}{2\cdot 3}\right)$$

$$\frac{1}{2\cdot 3\cdot 4}=\frac{1}{2}\left(\frac{1}{2\cdot 3}-\frac{1}{3\cdot 4}\right)$$

$$\cdots\cdots\cdots\cdots\cdots\cdots\cdots\cdots\cdots\cdots$$

$$\cdots\cdots\cdots\cdots\cdots\cdots\cdots\cdots\cdots$$

$$\text{Sum}=\frac{1}{2}\left[\left(\frac{1}{1\cdot 2}-\frac{1}{2\cdot 3}\right)+\left(\frac{1}{2\cdot 3}-\frac{1}{3\cdot 4}\right)+\cdots\right]$$

$$=\frac{1}{2}\left(\frac{1}{2}\right)$$

$$=\frac{1}{4}$$

# Proof 9

$$\frac{1}{1\cdot 3\cdot 5}+\frac{1}{3\cdot 5\cdot 7}+\frac{1}{5\cdot 7\cdot 9}+\cdots=\frac{1}{12}$$

$$\frac{1}{1\cdot 3\cdot 5}=\frac{1}{4}\left(\frac{1}{1\cdot 3}-\frac{1}{3\cdot 5}\right)$$

$$\frac{1}{3\cdot 5\cdot 7}=\frac{1}{4}\left(\frac{1}{3\cdot 5}-\frac{1}{5\cdot 7}\right)$$

..............................

..............................

$$\text{Sum}=\frac{1}{4}\left[\left(\frac{1}{1\cdot 3}-\frac{1}{3\cdot 5}\right)+\left(\frac{1}{3\cdot 5}-\frac{1}{5\cdot 7}\right)+\cdots\right]$$

$$=\frac{1}{4}\left(\frac{1}{3}\right)$$

$$=\frac{1}{12}$$

# Proof 10

$$\frac{1}{1 \cdot 2 \cdot 3} + \frac{2}{2 \cdot 3 \cdot 4} + \frac{3}{3 \cdot 4 \cdot 5} + \cdots = \frac{1}{2}$$

Although this series appears to be a new one, closer inspection shows that it is really identical to an earlier one:

$$\frac{1}{1 \cdot 2 \cdot 3} = \frac{1}{2 \cdot 3}$$

$$\frac{2}{2 \cdot 3 \cdot 4} = \frac{1}{3 \cdot 4}$$

$$\frac{3}{3 \cdot 4 \cdot 5} = \frac{1}{4 \cdot 5}$$

. . . . . . . . . . . . . . . . .

$$\text{Sum} = \frac{1}{2 \cdot 3} + \frac{1}{3 \cdot 4} + \frac{1}{4 \cdot 5} + \cdots$$

$$= [1] - \frac{1}{1 \cdot 2} \qquad \text{from Proof 1}$$

$$= \frac{1}{2}$$

Great minds in mathematics often find patterns in new problems that enable them to use solutions from previous problems. Such creativity is rare and often yields dividends as we shall see in the works of Euler and other geniuses. The above series is a trivial example of such insight.

# Proof 11

$$\frac{1}{1 \cdot 3 \cdot 5} + \frac{2}{3 \cdot 5 \cdot 7} + \frac{3}{5 \cdot 7 \cdot 9} + \cdots = \frac{1}{8}$$

Although formally similar in structure to the previous equation, this is now a new series because there is no simple cancellation to transform it into simpler terms. The $n$th term can be factorised into the sum of two simpler terms:

$$\frac{n}{(2n-1)(2n+1)(2n+3)} = \frac{1}{8}\left[\frac{1}{(2n-1)(2n+1)} + \frac{3}{(2n+1)(2n+3)}\right]$$

$\displaystyle\sum_{1}^{\infty}\frac{1}{(2n-1)(2n+1)}$ gives the series $\dfrac{1}{1 \cdot 3} + \dfrac{1}{3 \cdot 5} + \dfrac{1}{5 \cdot 7} + \cdots$

and sums to $\dfrac{1}{2}$ (from Proof 5).

Similarly,

$\displaystyle\sum_{1}^{\infty}\frac{1}{(2n+1)(2n+3)}$ gives the series $\dfrac{1}{3 \cdot 5} + \dfrac{1}{5 \cdot 7} + \dfrac{1}{7 \cdot 9} + \cdots$

and sums to $\left(\dfrac{1}{2} - \dfrac{1}{1 \cdot 3}\right)$ i.e., $\dfrac{1}{6}$

Therefore the sum of the series is

$$\frac{1}{8}\left[\frac{1}{2} + \frac{3}{6}\right]$$

$$= \frac{1}{8}$$

# Proof 12

$$\frac{1}{1}+\frac{1}{2}+\frac{1}{4}+\frac{1}{8}+\frac{1}{16}+\frac{1}{32}+\cdots=2$$

$$1=\frac{1}{2}+\frac{1}{2}$$

$$=\frac{1}{2}+\left(\frac{1}{4}+\frac{1}{4}\right)$$

$$=\frac{1}{2}+\frac{1}{4}+\left(\frac{1}{8}+\frac{1}{8}\right)$$

$$=\frac{1}{2}+\frac{1}{4}+\frac{1}{8}+\left(\frac{1}{16}+\frac{1}{16}\right)$$

..........................................

$$\therefore \quad \frac{1}{1}+\frac{1}{2}+\frac{1}{4}+\frac{1}{8}+\frac{1}{16}+\frac{1}{32}+\cdots$$

$$=2$$

The "Geometric Series" is of great interest in the history of mathematics, and served as our introduction to convergent series. As is often true in mathematics, writing the series in a different form opens up a completely new perspective with tremendous implications:

$$\frac{1}{2^0}+\frac{1}{2^1}+\frac{1}{2^2}+\frac{1}{2^3}+\cdots=2$$

This "Geometric Series" is a special case of a class of infinite series better known as geometric progressions.

Consider the general equation:

$$\sum_{1}^{n} ar^{i-1} = a + ar + ar^2 + ar^3 + \cdots + ar^{n-1} \left( \sum_{1}^{n} ar^{i-1} \text{ is abbreviated to } \sum_{1}^{n} \text{ below} \right)$$

where $a$ is the first term, and $r$ is the common ratio between two consecutive terms.

Multiplying both sides of the equation by $r$:

$$r\sum_{1}^{n} = ar + ar^2 + ar^3 + ar^4 + \cdots + ar^n$$

Subtracting the first equation from the second:

$$r\sum_{1}^{n} - \sum_{1}^{n} = ar^n - a$$

$$\therefore \quad (r-1)\sum_{1}^{n} = a(r^n - 1)$$

$$\therefore \quad \sum_{1}^{n} = \frac{a(r^n - 1)}{(r - 1)}$$

or

$$\sum_{1}^{n} = \frac{a(1 - r^n)}{(1 - r)}$$

For $-1 < r < 1$,

$$r^n \to 0 \text{ as } n \to \infty$$

$$\therefore \quad \sum_{1}^{\infty} = \frac{a}{1 - r}$$

Applying this general equation back to the Geometric Series:

$$\frac{1}{2^0} + \frac{1}{2^1} + \frac{1}{2^2} + \frac{1}{2^3} + \frac{1}{2^4} + \cdots$$

$a = 1$, and $r = \dfrac{1}{2}$

$$\therefore \quad \sum_{1}^{\infty} = \frac{1}{1 - \dfrac{1}{2}} = \frac{1}{\dfrac{1}{2}}$$

$$= 2$$

Earlier on we saw that the sum of the Harmonic Series tends to infinity:

$$\frac{1}{1} + \frac{1}{2} + \frac{1}{3} + \frac{1}{4} + \cdots \longrightarrow \infty$$

What about the series

$$1 + \left(\frac{9}{10}\right)^1 + \left(\frac{9}{10}\right)^2 + \left(\frac{9}{10}\right)^3 + \left(\frac{9}{10}\right)^4 + \cdots ?$$

Intuitively we can see that

$$\left(\frac{9}{10}\right)^1 \qquad\qquad > \frac{1}{2}$$

$$\left(\frac{9}{10}\right)^2 = \frac{81}{100} \quad > \frac{1}{3}$$

$$\left(\frac{9}{10}\right)^3 = \frac{729}{1000} > \frac{1}{4}$$

So we may be tempted to infer that since it appears that the individual terms of this new series are greater than the corresponding terms in the Harmonic Series, this new series would also tend to infinity for its sum.

However, we note that this new series is a member of the geometric series class, with

$$a = 1 \text{ and } r = \frac{9}{10}.$$

Hence the sum is:

$$\sum_{1}^{\infty} = \frac{1}{1 - \frac{9}{10}} = \frac{1}{\frac{1}{10}}$$

$$= 10$$

Surprise, surprise! The series sums to only a relatively small constant, namely 10. How does a series beginning with such large fractions sum to 10 whereas the Harmonic Series tends to infinity for its sum?

Take an even more extreme example:

$$1 + \frac{999,999}{1,000,000} + \left(\frac{999,999}{1,000,000}\right)^2 + \left(\frac{999,999}{1,000,000}\right)^3 \cdots$$

Does the sum of these "very big" terms now tend to infnity? Again the answer is "No".

It's a geometric series:

$$a = 1 \text{ and } r = \frac{999,999}{1,000,000}$$

$$\therefore \sum_{1}^{\infty} = \frac{1}{1 - \frac{999,999}{1,000,000}} = \frac{1}{\frac{1}{1,000,000}}$$

$$= 1,000,000$$

Superficially, the last two series appear to have second and subsequent terms which are much larger than the corresponding terms in the Harmonic Series. Yet both of them sum to constants. This is because for terms further down the series, as $n$ gets larger and larger, the terms of these two series become smaller much faster than their corresponding terms in the Harmonic Series, resulting in their finite sums.

This is another good illustration of many problems in mathematics, where a superficial analysis can lead to serious erroneous conclusions.

To illustrate further the power of the general equation, let us look at the series:

$$\frac{9}{10} + \frac{9}{10^2} + \frac{9}{10^3} + \cdots$$

$$a = \frac{9}{10}, \ r = \frac{1}{10}$$

$$\therefore \ \sum_{1}^{\infty} = \frac{\frac{9}{10}}{1 - \frac{1}{10}} = \frac{9}{10} \cdot \frac{10}{9}$$

$$= 1$$

Surprising, isn't it?

This was the series that my granddaughter Rebecca refused to believe when I told her that it sums to 1 — until I showed her "the visual proof":

$$\frac{9}{10} = 0.9$$

$$\frac{9}{10^2} = \frac{9}{100} = 0.09$$

$$\frac{9}{10^3} = \frac{9}{1000} = 0.009$$

$$\frac{9}{10^4} = \frac{9}{10\,000} = 0.0009$$

$$\cdots\cdots\cdots\cdots\cdots\cdots\cdots$$

$$\cdots\cdots\cdots\cdots\cdots\cdots\cdots$$

$$\therefore \ \sum_{1}^{\infty} = 0.9999\ldots$$

$$= 1$$

Seeing that

$$\frac{1}{2^1} + \frac{1}{2^2} + \frac{1}{2^3} + \frac{1}{2^4} + \cdots = 1$$

$$\frac{9}{10^1} + \frac{9}{10^2} + \frac{9}{10^3} + \frac{9}{10^4} + \cdots = 1,$$

do you see the general pattern for summing to 1?

If so, can you come up with your own geometric series which sums to 1? (Pause for a while and see if you can be a creative mathematician. Give yourself time to think, before you turn the page and look at the answer.)

Now turn the page and see some other geometric series which sum to 1.

$$\frac{2}{3^1}+\frac{2}{3^2}+\frac{2}{3^3}+\frac{2}{3^4}+\cdots \qquad =1$$

$$\frac{3}{4^1}+\frac{3}{4^2}+\frac{3}{4^3}+\frac{3}{4^4}+\cdots \qquad =1$$

$$\frac{n}{(n+1)}+\frac{n}{(n+1)^2}+\frac{n}{(n+1)^3}+\cdots =1$$

$$\sum_{1}^{\infty}=\frac{a}{1-r} \quad \text{where} \quad a=\frac{n}{(n+1)},\ r=\frac{1}{(n+1)}$$

$$=\frac{\dfrac{n}{(n+1)}}{1-\dfrac{1}{(n+1)}}$$

$$=\frac{n}{(n+1)}\cdot\frac{(n+1)}{n}$$

$$=1$$

Amazing, isn't it?

The geometric series is one of only a few classes of infinite series that has a general equation for summing. Most other series require a genius such as Euler to come up with a new perspective, and then a proof.

For the sister series with alternating plus and minus terms:

$$\frac{3}{2^1} - \frac{3}{2^2} + \frac{3}{2^3} - \frac{3}{2^4} + \cdots \qquad = 1$$

$$\cdots\cdots\cdots\cdots\cdots\cdots$$

$$\frac{n+1}{n^1} - \frac{n+1}{n^2} + \frac{n+1}{n^3} - \frac{n+1}{n^4} + \cdots = 1$$

$$\text{Sum} = \frac{a}{1-r} = \left(\frac{n+1}{n}\right)\frac{1}{\left(1-\left(\frac{-1}{n}\right)\right)} \quad \text{where} \quad a = \frac{n+1}{n}; \ r = \left(-\frac{1}{n}\right)$$

$$= \left(\frac{n+1}{n}\right)\left(\frac{n}{n+1}\right)$$

$$= 1$$

For the sister series of "the Geometric Series" with alternating plus and minus terms:

$$\frac{1}{1} - \frac{1}{2} + \frac{1}{4} - \frac{1}{8} + \frac{1}{16} - \cdots$$

$$\text{Sum} = \frac{a}{1-r} \quad \text{where} \quad a = 1, \ r = \left(-\frac{1}{2}\right)$$

$$= \frac{1}{1-\left(-\frac{1}{2}\right)}$$

$$= \frac{1}{\frac{3}{2}}$$

$$= \frac{2}{3}$$

# Proof 13

$$\frac{1}{2^1} + \frac{2}{2^2} + \frac{3}{2^3} + \frac{4}{2^4} + \cdots = 2$$

From "the Geometric Series", we get:

$$\frac{1}{2^1} + \frac{1}{2^2} + \frac{1}{2^3} + \frac{1}{2^4} + \frac{1}{2^5} + \cdots = 1$$

$$\frac{1}{2^1} + \frac{2}{2^2} + \frac{3}{2^3} + \frac{4}{2^4} + \frac{5}{2^5} + \cdots$$

$$= \begin{cases} \dfrac{1}{2^1} + \dfrac{1}{2^2} + \dfrac{1}{2^3} + \dfrac{1}{2^4} + \dfrac{1}{2^5} + \cdots = 1 & = 1 \\[2ex] \phantom{\dfrac{1}{2^1} +} \dfrac{1}{2^2} + \dfrac{1}{2^3} + \dfrac{1}{2^4} + \dfrac{1}{2^5} + \cdots = 1 - \dfrac{1}{2} & = \dfrac{1}{2} \\[2ex] \phantom{\dfrac{1}{2^1} + \dfrac{1}{2^2} +} \dfrac{1}{2^3} + \dfrac{1}{2^4} + \dfrac{1}{2^5} + \cdots = 1 - \dfrac{1}{2} - \dfrac{1}{4} = \dfrac{1}{4} \\[2ex] \cdots\cdots\cdots\cdots\cdots\cdots\cdots\cdots\cdots\cdots\cdots \\ \cdots\cdots\cdots\cdots\cdots\cdots\cdots\cdots\cdots\cdots\cdots \end{cases}$$

$$\overline{\phantom{xxxxxxxxxxxxx}} = 2$$

Can you follow the vertical addition on the extreme right? It is the same "Geometric Series"?

Can you find a simpler or more beautiful proof?

# Proof 14

$$\boxed{\frac{1}{1}+\frac{1}{2}+\frac{1}{3}+\frac{1}{4}+\frac{1}{5}+\cdots \rightarrow \infty}$$

$$\frac{1}{1}+\frac{1}{2}+\frac{1}{3}+\frac{1}{4}+\frac{1}{5}+\cdots$$

$$\text{Sum} = \frac{1}{1}+\frac{1}{2}+\left(\frac{1}{3}+\frac{1}{4}\right)+\left(\frac{1}{5}+\frac{1}{6}+\frac{1}{7}+\frac{1}{8}\right)+\cdots$$

Since $\frac{1}{3}>\frac{1}{4}$ (the sign $>$ means "greater than")

$$\left(\frac{1}{3}+\frac{1}{4}\right)>\left(\frac{1}{4}+\frac{1}{4}\right)=\frac{1}{2}$$

Similarly,

$$\frac{1}{5}>\frac{1}{6}>\frac{1}{7}>\frac{1}{8}$$

$$\therefore \quad \left(\frac{1}{5}+\frac{1}{6}+\frac{1}{7}+\frac{1}{8}\right)>\left(\frac{1}{8}+\frac{1}{8}+\frac{1}{8}+\frac{1}{8}\right)=\frac{1}{2}$$

$$\therefore \quad \text{Sum} > \frac{1}{1}+\frac{1}{2}+\left(\frac{1}{2}\right)+\left(\frac{1}{2}\right)+\cdots$$

By taking increasing number of terms to add up to a partial sum that is greater than, $\frac{1}{2}$, we can make the sum greater than an infinite number of $\frac{1}{2}$'s. Hence the sum of "the Harmonic Series" tends to infinity.

Using the same method, we can prove that the half "Harmonic Series" wih odd integers only and that with even integers only also give sums which tend to infinity:

$$\frac{1}{1} + \frac{1}{3} + \frac{1}{5} + \frac{1}{7} + \frac{1}{9} + \cdots$$

$$= \frac{1}{1} + \frac{1}{3} + \left(\frac{1}{5} + \frac{1}{7}\right) + \left(\frac{1}{9} + \frac{1}{11} + \frac{1}{13} + \frac{1}{15}\right) + \cdots$$

$$> \frac{1}{1} + \frac{1}{3} + \left(\frac{1}{3}\right) + \left(\frac{1}{3}\right) + \cdots$$

$$\to \infty$$

# Proof 15

$$\boxed{\frac{1}{1} - \frac{1}{2} + \frac{1}{3} - \frac{1}{4} + \frac{1}{5} - \cdots = \log_{natural} 2}$$

From "Elementary Calculus", we know that

$$\frac{d}{dx}(\ln x) = \frac{1}{x}$$

$$\therefore \quad \frac{d}{dx}(\ln(1+x)) = \frac{1}{1+x}$$

Conversely, by integration,

$$\int_1^\infty \frac{1}{x} \cdot dx = \ln x \qquad \left( \int_1^\infty \text{ is abbreviated to } \int \text{ in the rest of this book} \right)$$

$$\int \frac{1}{1+x} \cdot dx = \ln(1+x)$$

By normal division,

$$\frac{1}{1+x} = 1 - x + x^2 - x^3 + x^4 - \cdots$$

$$\ln(1+x) = \int \frac{1}{1+x} \cdot dx = x - \frac{x^2}{2} + \frac{x^3}{3} - \frac{x^4}{4} + \frac{x^5}{5} - \cdots$$

Let $x = 1$

$$\therefore \quad \ln(1+1) = \frac{1}{1} - \frac{1}{2} + \frac{1}{3} - \frac{1}{4} + \frac{1}{5} - \cdots$$

$$\ln 2 = \frac{1}{1} - \frac{1}{2} + \frac{1}{3} - \frac{1}{4} + \frac{1}{5} - \cdots$$

# Proof 16

$$\frac{1}{1 \cdot 2} + \frac{2}{2 \cdot 3} + \frac{3}{3 \cdot 4} + \frac{4}{4 \cdot 5} + \cdots \to \infty$$

$$\frac{1}{1 \cdot 2} + \frac{2}{2 \cdot 3} + \frac{3}{3 \cdot 4} + \frac{4}{4 \cdot 5} + \cdots$$

$$= \frac{1}{2} + \frac{1}{3} + \frac{1}{4} + \frac{1}{5} + \cdots$$

$$\to \infty$$

since the series is "the Harmonic Series" less the first term 1.

# Proof 17

$$\frac{1}{1\cdot 2}+\frac{2}{3\cdot 4}+\frac{3}{5\cdot 6}+\frac{4}{7\cdot 8}+\cdots \to \infty$$

$$\frac{1}{1\cdot 2}+\frac{2}{3\cdot 4}+\frac{3}{5\cdot 6}+\frac{4}{7\cdot 8}+\cdots$$

$$=\frac{1}{2}+\frac{1}{3\cdot 2}+\frac{1}{5\cdot 2}+\frac{1}{7\cdot 2}+\cdots$$

$$=\frac{1}{2}\left(1+\frac{1}{3}+\frac{1}{5}+\frac{1}{7}+\cdots\right)$$

$$\to \infty$$

since the series in bracket is "the half Harmonic Series" with odd integers only, with sum tending to infinity.

# Proof 18

$$\frac{1}{1\cdot3}+\frac{2}{3\cdot5}+\frac{3}{5\cdot7}+\frac{4}{7\cdot9}+\cdots\rightarrow\infty$$

$$\frac{1}{1\cdot3}=\frac{1}{2}\left(\frac{1}{1}-\frac{1}{3}\right)$$

$$\frac{2}{3\cdot5}=\frac{2}{2}\left(\frac{1}{3}-\frac{1}{5}\right)$$

$$\frac{3}{5\cdot7}=\frac{3}{2}\left(\frac{1}{5}-\frac{1}{7}\right)$$

$$\text{Sum}=\frac{1}{2}\left(\frac{1}{1}-\frac{1}{3}\right)+\frac{2}{2}\left(\frac{1}{3}-\frac{1}{5}\right)+\frac{3}{2}\left(\frac{1}{5}-\frac{1}{7}\right)+\cdots$$

$$=\frac{1}{2}\left(\frac{1}{1}+\frac{1}{3}+\frac{1}{5}+\frac{1}{7}+\cdots\right)$$

$$\rightarrow\infty$$

since the sum of "the half Harmonic Series" (in brackets) tends to infinity.

It is interesting to note that though the series:

$$\frac{1}{1\cdot2}+\frac{1}{3\cdot4}+\frac{1}{5\cdot6}+\frac{1}{7\cdot8}+\cdots \text{ sums to } \ln 2,$$

and the series:

$$\frac{1}{1\cdot3}+\frac{1}{3\cdot5}+\frac{1}{5\cdot7}+\frac{1}{7\cdot9}+\cdots \text{ sums to } \frac{1}{2},$$

when we change the numerators from 1 for all the terms to 1, 2, 3, 4 ... $n$ progressively for the terms, both the series become identical and sum to half of "the half Harmonic Series with odd integers only"!

# Proof 19

$$\frac{1}{1\cdot 3}+\frac{2}{5\cdot 7}+\frac{3}{9\cdot 11}+\frac{4}{13\cdot 15}+\cdots \rightarrow \infty$$

We will use an indirect approach to provide the proof for the above series. Indirect approaches are often used in mathematics.

Let us use a sister series:

$$\frac{1}{1\cdot 5}+\frac{2}{5\cdot 9}+\frac{3}{9\cdot 13}+\frac{4}{13\cdot 17}+\cdots$$

$$\frac{1}{1\cdot 5}=\frac{1}{4}\left(\frac{1}{1}-\frac{1}{5}\right)$$

$$\frac{2}{5\cdot 9}=\frac{2}{4}\left(\frac{1}{5}-\frac{1}{9}\right)$$

$$\frac{3}{9\cdot 13}=\frac{3}{4}\left(\frac{1}{9}-\frac{1}{13}\right)$$

$$\text{Sum}=\frac{1}{4}\left(\frac{1}{1}-\frac{1}{5}\right)+\frac{2}{4}\left(\frac{1}{5}-\frac{1}{9}\right)+\frac{3}{4}\left(\frac{1}{9}-\frac{1}{13}\right)+\cdots$$

$$=\frac{1}{4}\left[\frac{1}{1}+\frac{1}{5}+\frac{1}{9}+\frac{1}{13}+\cdots\right]$$

$$\rightarrow \infty$$

since the series in brackets is a quarter subset of the Harmonic Series with a sum that tends to infinity.

If we look at the two series:

$$\frac{1}{1\cdot 3}+\frac{2}{5\cdot 7}+\frac{3}{9\cdot 11}+\frac{4}{13\cdot 15}+\cdots \quad \text{and}$$

$$\frac{1}{1\cdot 5}+\frac{2}{5\cdot 9}+\frac{3}{9\cdot 13}+\frac{4}{13\cdot 17}+\cdots$$

we can see that *every* term in the first series is larger than the corresponding term in the second series. Since we have proven that the sum of the second series tends to infinity, we can also conclude that the sum of the first series tends to infinity.

# Proof 20

$$\frac{1}{1\cdot 2}+\frac{2}{2\cdot 3}+\frac{3}{3\cdot 4}+\frac{4}{4\cdot 5}+\frac{5}{5\cdot 6}+\cdots \rightarrow \infty$$

$$\frac{1}{2\cdot 3}+\frac{2}{3\cdot 4}+\frac{3}{4\cdot 5}+\frac{4}{5\cdot 6}+\cdots \rightarrow \infty$$

$$\frac{1}{3\cdot 4}+\frac{2}{4\cdot 5}+\frac{3}{5\cdot 6}+\cdots \rightarrow \infty$$

$$\cdots\cdots\cdots\cdots\cdots\cdots\cdots\cdots\cdots\cdots\cdots\cdots$$

$$\frac{1}{1,000,000\cdot 1,000,001}+\frac{2}{1,000,001\cdot 1,000,002}\rightarrow \infty$$

$$\frac{1}{2\cdot 3}+\frac{2}{3\cdot 4}+\frac{3}{4\cdot 5}+\cdots$$

$$=\left(\frac{1}{2}-\frac{1}{3}\right)+2\left(\frac{1}{3}-\frac{1}{4}\right)+3\left(\frac{1}{4}-\frac{1}{5}\right)+\cdots$$

$$=\frac{1}{2}+\frac{1}{3}+\frac{1}{4}+\frac{1}{5}+\cdots$$

$$\rightarrow \infty$$

since it is "the Harmonic Series" less 1.

Similarly

$$\frac{1}{n(n+1)}+\frac{2}{(n+1)(n+2)}+\cdots$$

$$=\left[\left(\frac{1}{n}-\frac{1}{n+1}\right)\right]+2\left[\left(\frac{1}{(n+1)}-\frac{1}{(n+2)}\right)\right]$$

$$=\frac{1}{n}+\frac{1}{n+1}+\frac{1}{n+2}+\cdots$$

$$\rightarrow \infty$$

# Proof 21

$$\frac{1}{1\cdot 2}+\frac{1}{2\cdot 3}+\frac{1}{3\cdot 4}+\frac{1}{4\cdot 5}+\cdots = 1$$

$$\frac{1}{1\cdot 2}+\frac{2}{2\cdot 3}+\frac{3}{3\cdot 4}+\frac{4}{4\cdot 5}+\cdots \rightarrow \infty$$

$$\frac{1}{1\cdot 3}+\frac{2}{3\cdot 5}+\frac{3}{5\cdot 7}+\frac{4}{7\cdot 9}+\cdots \rightarrow \infty$$

$$\frac{1}{1\cdot 4}+\frac{2}{4\cdot 7}+\frac{3}{7\cdot 10}+\cdots \qquad \rightarrow \infty$$

$$\cdots\cdots\cdots\cdots\cdots\cdots\cdots\cdots\cdots\cdots\cdots\cdots\cdots$$

$$\cdots\cdots\cdots\cdots\cdots\cdots\cdots\cdots\cdots\cdots\cdots\cdots\cdots$$

$$\frac{1}{1\cdot n}+\frac{2}{n(2n-1)}+\frac{3}{(2n-1)(3n-2)}+\cdots \rightarrow \infty$$

For the general equation:

$$\frac{1}{1\cdot n}=\frac{1}{(n-1)}\left(\frac{1}{1}-\frac{1}{n}\right)$$

$$\frac{2}{n(2n-1)}=\frac{2}{(n-1)}\left(\frac{1}{n}-\frac{1}{(2n-1)}\right)$$

$$\cdots\cdots\cdots\cdots\cdots\cdots\cdots\cdots\cdots\cdots\cdots\cdots$$

$$\cdots\cdots\cdots\cdots\cdots\cdots\cdots\cdots\cdots\cdots\cdots\cdots$$

$$\therefore \ \text{Sum}=\frac{1}{(n-1)}\left[\left(\left(\frac{1}{1}-\frac{1}{n}\right)+2\left(\frac{1}{n}-\frac{1}{(2n-1)}\right)+\cdots\right)\right]$$

$$=\frac{1}{(n-1)}\left[\frac{1}{1}+\frac{1}{n}+\frac{1}{(2n-1)}+\frac{1}{(3n-2)}+\cdots\right]$$

$$\rightarrow \infty$$

since the series in brackets is a subset of the "Harmonic Series".

# A Note of Caution

In the context of infinite series, the basic operations of arithmetic — namely, addition, subtraction, multiplication and division are not permitted, with the exception of the infinite series which are "absolutely convergent".

An infinite series is "divergent" if its sum tends to infinity as $n$ tends to infinity. The series is "convergent" if its sum tends to a constant when $n$ tends to infinity.

"Convergent" series can be either "conditionally convergent" or "absolutely convergent". "Conditionally convergent" series are generally made up of terms with alternating signs (positive followed by negative). Should the negative terms be changed to positive, the series ceases to be "convergent" and becomes "divergent". "Absolutely convergent" series can consist of series with all positive terms or series with both positive and negative term. In the case of the latter, the series continues to be "convergent" even when the negative terms are made positive.

The beautiful Liebniz-Gregory series which sums to $\frac{\pi}{4}$ is "convergent". It is made up of alternating terms.

$$\frac{1}{1} - \frac{1}{3} + \frac{1}{5} - \frac{1}{7} + \frac{1}{9} - \cdots = \frac{\pi}{4}$$

When the negative terms are made positive, the series becomes the "half Harmonic Series":

$$\frac{1}{1} + \frac{1}{3} + \frac{1}{5} + \frac{1}{7} + \frac{1}{9} + \cdots \to \infty$$

which is divergent. Hence the Leibniz-Gregory Series is "conditionally convergent". Therefore we cannot rearrange the terms. Otherwise we end up with "meaningless" mathematics.

An example of "meaningless" mathematics:

$$\frac{1}{1} - \frac{1}{3} + \frac{1}{5} - \frac{1}{7} + \frac{1}{9} - \frac{1}{11} + \cdots$$

$$= \left( \frac{1}{1} + \frac{1}{5} + \frac{1}{9} + \cdots \right) - \left( \frac{1}{3} + \frac{1}{7} + \frac{1}{11} + \cdots \right)$$

$$= (\infty) - (\infty)$$

which is meaningless.

Another example of meaningless mathematics:

$$\frac{1}{1} - \frac{1}{2} + \frac{1}{3} - \frac{1}{4} + \frac{1}{5} - \frac{1}{6} + \frac{1}{7} - \cdots = \ln 2$$

Multiplying both sides of the equation by $\frac{1}{2}$:

$$\frac{1}{2} - \frac{1}{4} + \frac{1}{6} - \frac{1}{8} + \frac{1}{10} - \frac{1}{12} + \frac{1}{14} - \cdots = \frac{1}{2} \ln 2$$

Adding the terms of both series:

$$\frac{1}{1} + \frac{1}{3} - \frac{2}{4} + \frac{1}{5} - \frac{2}{8} + \frac{1}{7} - \cdots = \frac{3}{2} \ln 2$$

Rearranging the terms:

$$\frac{1}{1} - \frac{1}{2} + \frac{1}{3} - \frac{1}{4} + \frac{1}{5} - \frac{1}{6} + \frac{1}{7} - \cdots = \frac{3}{2} \ln 2$$

We obtain the original series, but it is now equal to $\frac{3}{2} \ln 2$ — a totally erroneous conclusion.

Hence the emphasis that arithmetical operations are valid only for "absolutely convergent" series.

# Less Easy Proofs

There is divinity in odd numbers

**William Shakespheare** (1564–1616)

§

Each problem that I solve
becomes a rule
which serves afterwards
to solve other problems

**Rene Descartes** (1596–1650)

§

# Proof 22

$$1+\frac{1}{1!}+\frac{1}{2!}+\frac{1}{3!}+\frac{1}{4!}+\cdots=e$$

Let us look at "the Geometric Series" and the $(e-1)$ series.

$$1+\frac{1}{2}+\frac{1}{4}+\frac{1}{8}+\frac{1}{16}+\cdots \quad = 2$$

$$\frac{1}{1!}+\frac{1}{2!}+\frac{1}{3!}+\frac{1}{4!}+\frac{1}{5!}+\cdots = (e-1)$$

Comparing the third and subsequent terms in the two series:

$$\frac{1}{3!}=\frac{1}{1\times2\times3}<\frac{1}{4}$$

$$\frac{1}{4!}=\frac{1}{1\times2\times3\times4}<\frac{1}{8}$$

$$\frac{1}{5!}=\frac{1}{1\times2\times3\times4\times5}<\frac{1}{16},$$

$$\cdots\cdots\cdots\cdots\cdots\cdots\cdots\cdots\cdots$$

$$\cdots\cdots\cdots\cdots\cdots\cdots\cdots\cdots\cdots$$

it is obvious that the sum of the $(e-1)$ series is less than 2. Therefore e must be less than 3. Simple addition of the terms in the e series gives the value of e to be 2.7182818284 ...

# Proof 23

$$1 + \frac{1}{1!} + \frac{1}{2!} + \frac{1}{3!} + \frac{1}{4!} + \cdots = e$$

$$\frac{1}{0!} + \frac{1}{1!} + \frac{1}{2!} + \frac{1}{3!} + \frac{1}{4!} + \cdots = e$$

e is normally written as 1 plus a series of reciprocals of the factorials of all the integers beginning with 1. The symbol:

$\sum_{1}^{\infty} f(n)$, sometimes, abbreviated to $\sum_{1}^{\infty}$, is the mathematician's abbreviation for the phrase "the sum of all terms for the function $f(n)$ going from $n = 1$ to $n$ tending to infinity". What usefulness in mathematical symbols! A simple (OK, not so simple) symbol for a concept that is quite a mouthful to describe in words.

Mathematicians excel in abbreviations and the use of Greek alphabets, and use them extensively — indeed so much so that such abbreviations and symbols create a major negative contributory factor to non-mathematicians' understanding of mathematics. Hence the "fear" of and "total incomprehension" of math by most people. The term $\sum_{1}^{\infty} \frac{1}{n!}$ combines four set of abbreviation:

1. $\sum$ is sum of all terms

2. $\sum_{1}^{\infty}$ is sum of all the terms from $n = 1$ to $n \rightarrow \infty$.

3. $n$ is any positive integer, and

4. $n!$ is "$n$ factorial", the product of all the integers from 1 to $n$ (i.e. $n! = 1 \times 2 \times 3 \times \ldots \times n$)

For the definition of factorial, it is important to bear in mind that contrary to expectation, 0! is 1 (Yes, one; not zero!). Hence the series can be written as in the second series above beginning with $\frac{1}{0!}$, or

$$\sum_{0}^{\infty} \frac{1}{n!} = e$$

This $\Sigma$ notation is extremely useful for the rest of this chapter, for the proofs for different e-series, and demonstrates the versatility of mathematical notations and abbreviations for mathematical manipulations. You must pay special attention to the sub- and superscript of the $\Sigma$ sign for they can vary and would give different results, e.g.,

$$\sum_{0}^{\infty} \text{ is not the same as } \sum_{1}^{\infty} \text{ or } \sum_{2}^{\infty} \text{ or } \sum_{0}^{n}$$

even for the same function. This is because different terms will be included in the summation.

Using the summation notation above, we can write the infinite series of e in a number of different ways:

$$e = 1 + \frac{1}{1!} + \frac{1}{2!} + \frac{1}{3!} + \frac{1}{4!} + \cdots$$

$$= 1 + \sum_{1}^{\infty} \frac{1}{n!} \qquad \qquad \therefore \ \sum_{1}^{\infty} \frac{1}{n!} = (e-1)$$

$$= 1 + 1 + \sum_{1}^{\infty} \frac{1}{(n+1)!} \qquad \qquad \therefore \ \sum_{1}^{\infty} \frac{1}{(n+1)!} = (e-2)$$

Also $\qquad e = \frac{1}{0!} + \frac{1}{1!} + \frac{1}{2!} + \frac{1}{3!} + \cdots$

$$= \sum_{1}^{\infty} \frac{1}{(n-1)!} \qquad \qquad \therefore \ \sum_{1}^{\infty} \frac{1}{(n-1)!} = e$$

e can also be written as a beautiful infinite nesting of increasing integers:

$$e = 1 + 1\left(1 + \frac{1}{2}\left(1 + \frac{1}{3}\left(1 + \frac{1}{4}\left(1 + \frac{1}{5}\cdots\right)\right)\right)\right)$$

$$= 1 + \frac{1}{1!} + \frac{1}{2!}\left(1 + \frac{1}{3}\left(1 + \frac{1}{4}\left(1 + \frac{1}{5}\cdots\right)\right)\right)$$

$$= 1 + \frac{1}{1!} + \frac{1}{2!} + \frac{1}{3!}\left(1 + \frac{1}{4}\left(1 + \frac{1}{5}\cdots\right)\right)$$

$$= 1 + \frac{1}{1!} + \frac{1}{2!} + \frac{1}{3!} + \frac{1}{4!}\left(1 + \frac{1}{5}\cdots\right)$$

Mathematically the function $e^x$ is unique, being the only function where the derivative and the integral are the same as the function itself:

i.e.
$$\frac{d}{dx}(e^x) = e^x$$

$$\int e^x dx = e^x + c \qquad (c \text{ is a constant})$$

It is useful to note that the e infinite series is "absolutely convergent" and the use of the arithmetical operations such as addition, subtraction and re-arranging of the individual terms is allowed. This enables us to prove the various summations of the different infinite series with relative ease.

# Proof 24

$$\frac{1}{1!} + \frac{2}{2!} + \frac{3}{3!} + \frac{4}{4!} + \cdots = e$$

$$\frac{1}{1!} + \frac{2}{2!} + \frac{3}{3!} + \frac{4}{4!} + \cdots$$

$$= 1 + \frac{2}{2 \cdot 1!} + \frac{3}{3 \cdot 2!} + \frac{4}{4 \cdot 3!} + \cdots$$

$$= 1 + \frac{1}{1!} + \frac{1}{2!} + \frac{1}{3!} + \cdots$$

$$= e$$

$$\therefore \quad \sum_{1}^{\infty} \frac{n}{n!} = e$$

# Proof 25

$$1 + \frac{3}{2!} + \frac{5}{4!} + \frac{7}{6!} + \cdots = e$$

Let us prove the above equation in two ways, the visual, and the formal way using $\Sigma$ notation.

$$\text{Sum} = 1 + \frac{3}{2!} + \frac{5}{4!} + \frac{7}{6!} + \cdots$$

$$= \begin{cases} 1 + \dfrac{2}{2!} + \dfrac{4}{4!} + \dfrac{6}{6!} + \cdots \\[2mm] + \dfrac{1}{2!} + \dfrac{1}{4!} + \dfrac{1}{6!} + \cdots \end{cases}$$

$$= \begin{cases} 1 + \dfrac{1}{1!} + \dfrac{1}{3!} + \dfrac{1}{5!} + \cdots \\[2mm] + \dfrac{1}{2!} + \dfrac{1}{4!} + \dfrac{1}{6!} + \cdots \end{cases}$$

$$= 1 + \frac{1}{1!} + \frac{1}{2!} + \frac{1}{3!} + \frac{1}{4!} + \frac{1}{5!} + \frac{1}{6!} + \cdots$$

$$= e$$

$$\text{Sum} = 1 + \frac{3}{2!} + \frac{5}{4!} + \frac{7}{6!} + \cdots$$

$$= 1 + \sum_{1}^{\infty} \frac{(2n+1)}{(2n)!}$$

$$= 1 + \sum_{1}^{\infty} \left[ \frac{2n}{(2n)!} + \frac{1}{(2n)!} \right]$$

$$= 1 + \sum_{1}^{\infty} \left[ \frac{1}{(2n-1)!} + \frac{1}{(2n)!} \right]$$

$$= 1 + \left( \frac{1}{1!} + \frac{1}{3!} + \frac{1}{5!} + \frac{1}{7!} + \cdots \right) + \left( \frac{1}{2!} + \frac{1}{4!} + \frac{1}{6!} + \cdots \right)$$

$$= 1 + \frac{1}{1!} + \frac{1}{2!} + \frac{1}{3!} + \frac{1}{4!} + \cdots$$

$$= e$$

Shall we work out the proof backwards for the fun of it, and also convince ourselves that our $\Sigma$ notation manipulation really works?

$$e = 1 + \frac{1}{1!} + \frac{1}{2!} + \frac{1}{3!} + \frac{1}{4!} + \cdots$$

$$= 1 + \left( \frac{1}{1!} + \frac{1}{3!} + \frac{1}{5!} + \cdots \right) + \left( \frac{1}{2!} + \frac{1}{4!} + \frac{1}{6!} + \cdots \right)$$

$$= 1 + \left( \frac{2}{2 \cdot 1!} + \frac{4}{4 \cdot 3!} + \frac{6}{6 \cdot 5!} + \cdots \right) + \left( \frac{1}{2!} + \frac{1}{4!} + \frac{1}{6!} + \cdots \right)$$

$$= 1 + \left( \frac{2}{2!} + \frac{4}{4!} + \frac{6}{6!} + \cdots \right) + \left( \frac{1}{2!} + \frac{1}{4!} + \frac{1}{6!} + \cdots \right)$$

$$= 1 + \frac{3}{2!} + \frac{5}{4!} + \frac{7}{6!} + \cdots$$

$$\therefore \quad \sum_{1}^{\infty} \frac{2n+1}{(2n)!} = (e-1)$$

$$\therefore \quad \sum_{0}^{\infty} \frac{2n+1}{(2n)!} = e$$

Remember that

$$\frac{2n}{(2n)!} = \frac{2n}{1 \times 2 \times 3 \times \cdots (2n-1)(2n)}$$

$$= \frac{1}{1 \times 2 \times 3 \times \cdots (2n-1)}$$

$$= \frac{1}{(2n-1)!} \qquad \left( \text{not } \frac{1}{1!} \right)$$

# Proof 26

$$\frac{1}{1!} + \frac{2}{3!} + \frac{3}{5!} + \frac{4}{7!} + \cdots = \frac{1}{2}e$$

$$\text{Sum} = \sum_{0}^{\infty} \frac{n+1}{(2n+1)!} = \frac{1}{2} \sum_{0}^{\infty} \frac{(2n+1+1)}{(2n+1)!}$$

$$= \frac{1}{2} \sum_{0}^{\infty} \left[ \frac{2n+1}{(2n+1)!} + \frac{1}{(2n+1)!} \right]$$

$$= \frac{1}{2} \sum_{0}^{\infty} \left[ \frac{1}{(2n)!} + \frac{1}{(2n+1)!} \right]$$

$$= \frac{1}{2} \left( \frac{1}{0!} + \frac{1}{2!} + \frac{1}{4!} + \frac{1}{6!} + \cdots + \frac{1}{1!} + \frac{1}{3!} + \frac{1}{5!} + \cdots \right)$$

$$= \frac{1}{2} \left( \frac{1}{0!} + \frac{1}{1!} + \frac{1}{2!} + \frac{1}{3!} + \frac{1}{4!} + \cdots \right)$$

$$= \frac{1}{2}e$$

$$\therefore \quad \sum_{0}^{\infty} \frac{n+1}{(2n+1)!} = \frac{1}{2}e$$

# Proof 27

$$\frac{2}{3!} + \frac{4}{5!} + \frac{6}{7!} + \frac{8}{9!} + \cdots = \frac{1}{e}$$

$$\text{Sum} = \sum_{1}^{\infty} \frac{2n}{(2n+1)!}$$

$$= \sum_{1}^{\infty} \left[ \frac{2n+1-1}{(2n+1)!} \right]$$

$$= \sum_{1}^{\infty} \left[ \frac{2n+1}{(2n+1)!} - \frac{1}{(2n+1)!} \right]$$

$$= \sum_{1}^{\infty} \left[ \frac{1}{(2n)!} - \frac{1}{(2n+1)!} \right]$$

$$= \left( \frac{1}{2!} + \frac{1}{4!} + \frac{1}{6!} + \cdots \right) - \left( \frac{1}{3} + \frac{1}{5} + \frac{1}{7} + \cdots \right)$$

$$= \frac{1}{2!} - \frac{1}{3!} + \frac{1}{4!} - \frac{1}{5!} + \cdots$$

$$= \left( 1 - \frac{1}{1!} \right) + \frac{1}{2!} - \frac{1}{3!} + \frac{1}{4!} - \cdots$$

$$= \frac{1}{e}$$

$$\therefore \quad \sum_{1}^{\infty} \frac{2n}{(2n+1)!} = \frac{1}{e}$$

# Proof 28

$$\frac{1}{2!} + \frac{2}{3!} + \frac{3}{4!} + \frac{4}{5!} + \cdots = 1$$

This is truly a stupendous, wondrous and mysterious series — with all factorials in the denominators but no e for its sum.

$$\text{Sum} = \sum_{1}^{\infty} \frac{n}{(n+1)!}$$

$$= \sum_{1}^{\infty} \left[ \frac{(n+1)-1}{(n+1)!} \right]$$

$$= \sum_{1}^{\infty} \left[ \frac{(n+1)}{(n+1)!} - \frac{1}{(n+1)!} \right]$$

$$= \sum_{1}^{\infty} \left[ \frac{1}{n!} - \frac{1}{(n+1)!} \right]$$

$$= (e-1) - (e-2) \qquad \text{from Proof 23}$$

$$= 1$$

$$\therefore \quad \sum_{1}^{\infty} \frac{n}{(n+1)!} = 1$$

We can also write:

$$(e-1) - (e-2)$$

$$= \left( \frac{1}{1!} + \frac{1}{2!} + \frac{1}{3!} + \frac{1}{4!} + \cdots \right) - \left( \frac{1}{2!} + \frac{1}{3!} + \frac{1}{4!} + \cdots \right)$$

$$= 1$$

Since it is such an elegant and beautiful series summing to 1, let us again prove the equation without the use of $\Sigma$ notation, for the fun of it.

$$\text{Sum} = \frac{1}{2!} + \frac{2}{3!} + \frac{3}{4!} + \frac{4}{5!} + \cdots$$

$$= \left\{ \begin{array}{l} \left( \dfrac{2}{2!} + \dfrac{3}{3!} + \dfrac{4}{4!} + \dfrac{5}{5!} + \cdots \right) \\[2ex] - \left( \dfrac{1}{2!} + \dfrac{1}{3!} + \dfrac{1}{4!} + \dfrac{1}{5!} + \cdots \right) \end{array} \right.$$

$$= \left\{ \begin{array}{l} \left( \dfrac{1}{1!} + \dfrac{1}{2!} + \dfrac{1}{3!} + \dfrac{1}{4!} + \cdots \right) \\[2ex] - \left( \dfrac{1}{2!} + \dfrac{1}{3!} + \dfrac{1}{4!} + \dfrac{1}{5!} + \cdots \right) \end{array} \right.$$

$$= 1$$

Earlier, we noted that this series has factorials in the denominators but does not sum to a constant involving e. If you think long and hard enough, you will remember that 1 is a very special number. It can be written as $e^0$, and so can be thought of as part of the e-family — $e^1$, $e^0$, $e^{-1}$ ... .

So the series with all the factorials in the denominator is part of the e-series, after all!

# Proof 29

$$1 + \frac{2}{1!} + \frac{3}{2!} + \frac{4}{3!} + \cdots = 2e$$

$$\text{Sum} = 1 + \sum_{1}^{\infty} \frac{n+1}{n!}$$

$$= 1 + \sum_{1}^{\infty} \left[ \frac{n}{n!} + \frac{1}{n!} \right]$$

$$= 1 + \sum_{1}^{\infty} \left[ \frac{1}{(n-1)!} + \frac{1}{n!} \right]$$

$$= 1 + e + (e-1) \qquad \text{from previous proofs}$$

$$= 2e$$

$$\therefore \quad \sum_{1}^{\infty} \frac{n+1}{n!} = (2e-1)$$

The series can also be written as:

$$\frac{1}{0!} + \frac{2}{1!} + \frac{3}{2!} + \frac{4}{3!} + \cdots$$

$$\therefore \quad \sum_{1}^{\infty} \frac{n}{(n-1)!} = 2e$$

# Proof 30

$$\frac{1^2}{1!} + \frac{2^2}{2!} + \frac{3^2}{3!} + \frac{4^2}{4!} + \cdots = 2e$$

$$\frac{1^2}{1!} + \frac{2^2}{2!} + \frac{3^2}{3!} + \frac{4^2}{4!} + \cdots$$

$$= \frac{1}{1} + \frac{2}{1!} + \frac{3}{2!} + \frac{4}{3!} + \cdots$$

$= 2e$      from previous proofs

$$\therefore \quad \sum_{1}^{\infty} \frac{n^2}{n!} = 2e$$

# Proof 31

$$\frac{1^2}{2!} + \frac{2^2}{3!} + \frac{3^2}{4!} + \frac{4^2}{5!} + \cdots = (e-1)$$

$$\text{Sum} = \sum_{1}^{\infty} \frac{n^2}{(n+1)!}$$

$$= \sum_{1}^{\infty} \left[ \frac{n^2 - 1 + 1}{(n+1)!} \right]$$

$$= \sum_{1}^{\infty} \left[ \frac{(n^2 - 1)}{(n+1)!} + \frac{1}{(n+1)!} \right]$$

$$= \sum_{1}^{\infty} \left[ \frac{n-1}{n!} + \frac{1}{(n+1)!} \right]$$

$$= \sum_{1}^{\infty} \left[ \frac{n}{n!} - \frac{1}{n!} + \frac{1}{(n+1)!} \right]$$

$$= e - (e-1) + (e-2) \qquad \text{from previous proofs}$$

$$= (e-1)$$

$$\therefore \quad \sum_{1}^{\infty} \frac{n^2}{(n+1)!} = (e-1)$$

# Proof 32

$$1 + \frac{2^2}{1!} + \frac{3^2}{2!} + \frac{4^2}{3!} + \cdots = 5e$$

$$\text{Sum} = 1 + \sum_1^\infty \frac{(n+1)^2}{n!}$$

$$= 1 + \sum_1^\infty \left[ \frac{n^2 + 2n + 1}{n!} \right]$$

$$= 1 + \sum_1^\infty \left[ \frac{n^2}{n!} + \frac{2}{(n-1)!} + \frac{1}{n!} \right]$$

$$= 1 + [2e + 2e + (e - 1)] \qquad \text{from previous proofs}$$

$$= 5e$$

$$\therefore \quad \sum_1^\infty \frac{(n+1)^2}{n!} = (5e - 1)$$

The series can also be rewritten as:

$$\frac{1^2}{0!} + \frac{2^2}{1!} + \frac{3^2}{2!} + \frac{4^2}{3!} + \cdots$$

$$\therefore \quad \sum_1^\infty \frac{n^2}{(n-1)!} = 5e$$

# Proof 33

$$\frac{1\cdot 3}{2!} + \frac{2\cdot 4}{3!} + \frac{3\cdot 5}{4!} + \cdots = (e+1)$$

$$\text{Sum} = \sum_{1}^{\infty} \frac{n(n+2)}{(n+1)!}$$

$$= \sum_{1}^{\infty} \frac{n^2 + 2n}{(n+1)!}$$

$$= \sum_{1}^{\infty} \frac{(n+1)^2 - 1}{(n+1)!}$$

$$= \sum_{1}^{\infty} \left[ \frac{n+1}{n!} - \frac{1}{(n+1)!} \right]$$

$$= \sum_{1}^{\infty} \left[ \frac{n}{n!} + \frac{1}{n!} - \frac{1}{(n+1)!} \right]$$

$$= e + (e-1) - (e-2) \qquad \text{from previous proofs}$$

$$= (e+1)$$

Alternatively we can re-write the series as

$$\left[ \frac{0\cdot 2}{1!} + \right] \frac{1\cdot 3}{2!} + \frac{2\cdot 4}{3!} + \frac{3\cdot 5}{4!} + \cdots$$

$$= \sum_{1}^{\infty} \frac{(n-1)(n+1)}{n!}$$

$$= \sum_{1}^{\infty} \frac{n^2 - 1}{n!}$$

$$= \sum_{1}^{\infty} \left[ \frac{n^2}{n!} - \frac{1}{n!} \right]$$

$$= 2e - (e-1) \qquad \text{from previous proofs}$$

$$= (e+1)$$

# Proof 34

$$\frac{1 \cdot 3}{2!} + \frac{3 \cdot 5}{4!} + \frac{5 \cdot 7}{6!} + \cdots = \frac{e^2 + 2e - 1}{2e}$$

$$\text{Sum} = \sum_{1}^{\infty} \frac{(2n-1)(2n+1)}{(2n)!}$$

$$= \sum_{1}^{\infty} \frac{2n(2n-1) + (2n-1)}{(2n)!}$$

$$= \sum_{1}^{\infty} \left[ \frac{2n(2n-1)}{(2n)!} + \frac{2n}{(2n)!} - \frac{1}{(2n)!} \right]$$

$$= \sum_{1}^{\infty} \left[ \frac{1}{(2n-2)!} + \frac{1}{(2n-1)!} - \frac{1}{(2n)!} \right]$$

Let's go back a little to work out these $2n$-type factorial terms:

$$e = 1 + \frac{1}{1!} + \frac{1}{2!} + \frac{1}{3!} + \frac{1}{4!} + \cdots$$

$$\frac{1}{e} = 1 - \frac{1}{1!} + \frac{1}{2!} - \frac{1}{3!} + \frac{1}{4!} - \cdots$$

$$e + \frac{1}{e} = 2 \left[ 1 + \frac{1}{2!} + \frac{1}{4!} + \cdots \right]$$

$$= 2 \sum_{1}^{\infty} \frac{1}{(2n-2)!}$$

$$\therefore \quad \sum_{1}^{\infty} \frac{1}{(2n-2)!} = \frac{1}{2} \left( e + \frac{1}{e} \right)$$

$$\frac{1}{e} = 1 - \frac{1}{1!} + \frac{1}{2!} - \frac{1}{3!} + \frac{1}{4!} - \cdots$$

$$= 1 - \left(\frac{1}{1!} + \frac{1}{3!} + \frac{1}{5!} + \cdots\right) + \left(\frac{1}{2!} + \frac{1}{4!} + \cdots\right)$$

$$= 1 - \sum_{1}^{\infty} \frac{1}{(2n-1)!} + \sum_{1}^{\infty} \frac{1}{(2n)!}$$

$$\therefore \quad \sum_{1}^{\infty} \left[\frac{1}{(2n-1)!} - \frac{1}{(2n)!}\right] = \left(1 - \frac{1}{e}\right)$$

Re-substituting back into the earlier calculations:

$$\text{Sum} = \sum_{1}^{\infty} \left[\frac{1}{(2n-2)!} + \frac{1}{(2n-1)!} - \frac{1}{(2n)!}\right]$$

$$= \frac{1}{2}\left(e + \frac{1}{e}\right) + \left(1 - \frac{1}{e}\right)$$

$$= \frac{e^2 + 1 + 2e - 2}{2e}$$

$$= \frac{e^2 + 2e - 1}{2e}$$

# Proof 35

$$\frac{2}{1!3} - \frac{3}{2!4} + \frac{4}{3!5} - \frac{5}{4!6} + \cdots = 3\left(\frac{1}{2} - \frac{1}{e}\right)$$

Multiplying both the numerator and the denominator by the same factors, viz. 2, for the first term, 3 for the second and so on, we get the series:

$$\text{Sum} = \frac{2^2}{3!} - \frac{3^2}{4!} + \frac{4^2}{5!} - \frac{5^2}{6!} + \cdots$$

We have shown earlier that the $\frac{n^2}{n!}$ series is "absolutely convergent" and sums to 2e; so by comparison of the corresponding terms, we can infer that the series $\frac{(n-1)^2}{n!}$ is also "absolutely convergent" since $\frac{(n-1)^2}{n!}$ is less than $\frac{n^2}{n!}$ for every term; similarly the $\frac{(n-1)^2}{n!}$ series with alternating terms is also absolutely convergent since it must be less than its sister series with all positive terms.

Hence we can rearrange the terms in the series!

$$\begin{aligned}
\text{Sum} &= \frac{2^2}{3!} - \frac{3^2}{4!} + \frac{4^2}{5!} - \frac{5^2}{6!} + \cdots \\
&= \left(\frac{2^2}{3!} + \frac{4^2}{5!} + \cdots\right) - \left(\frac{3^2}{4!} + \frac{5^2}{6!} + \cdots\right) \\
&= \sum_1^\infty \frac{(2n)^2}{(2n+1)!} - \sum_1^\infty \frac{(2n+1)^2}{(2n+2)!} \\
&= \sum_1^\infty \left[\frac{(2n+1)(2n)-(2n+1)+1}{(2n+1)!} - \frac{(2n+2)(2n+1)-(2n+2)+1}{(2n+2)!}\right] \\
&= \sum_1^\infty \left[\frac{1}{(2n-1)!} - \frac{1}{(2n)!} + \frac{1}{(2n+1)!} - \frac{1}{(2n)!} + \frac{1}{(2n+1)!} - \frac{1}{(2n+2)!}\right] \\
&= \sum_1^\infty \left[\frac{1}{(2n-1)!} - \frac{2}{(2n)!} + \frac{2}{(2n+1)!} - \frac{1}{(2n+2)!}\right]
\end{aligned}$$

Going back to basics:

$$e = 1 + \frac{1}{1!} + \frac{1}{2!} + \frac{1}{3!} + \frac{1}{4!} + \cdots$$

$$\frac{1}{e} = 1 - \frac{1}{1!} + \frac{1}{2!} - \frac{1}{3!} + \frac{1}{4!} - \cdots$$

$$e + \frac{1}{e} = 2\left(1 + \frac{1}{2!} + \frac{1}{4!} + \cdots\right)$$

$$\therefore \quad \sum_{1}^{\infty} \frac{1}{(2n)!} = \frac{1}{2}\left(e + \frac{1}{e} - 2\right)$$

$$\text{and } \sum_{1}^{\infty} \frac{1}{(2n+2)!} = \frac{1}{2}\left(e + \frac{1}{e} - 3\right)$$

$$e - \frac{1}{e} = 2\left(\frac{1}{1!} + \frac{1}{3!} + \frac{1}{5!} + \cdots\right)$$

$$\therefore \quad \sum_{1}^{\infty} \frac{1}{(2n-1)!} = \frac{1}{2}\left(e - \frac{1}{e}\right)$$

$$\text{and } \sum_{1}^{\infty} \frac{1}{(2n+1)!} = \frac{1}{2}\left(e - \frac{1}{e} - 2\right)$$

Re-substituting back into the calculation for sum:

$$\text{Sum} = \sum_{1}^{\infty} \left[ \frac{1}{(2n-1)!} - \frac{2}{(2n)!} + \frac{2}{(2n+1)!} - \frac{1}{(2n+2)!} \right]$$

$$= \frac{1}{2}\left(e - \frac{1}{e}\right) - \left(e + \frac{1}{e} - 2\right) + \left(e - \frac{1}{e} - 2\right) - \frac{1}{2}\left(e + \frac{1}{e} - 3\right)$$

$$= \frac{3}{2} - \frac{3}{e}$$

$$= 3\left(\frac{1}{2} - \frac{1}{e}\right) \qquad\qquad \text{whew!!!}$$

# Proof 36

$$\frac{1}{1!} + \frac{1+2}{2!} + \frac{1+2+3}{3!} + \frac{1+2+3+4}{4!} + \cdots = \frac{3e}{2}$$

The numerator is the sum of an arithmetical series:

$$1$$

$$1+2$$

$$1+2+3$$

$$S = 1+2+3+\ldots+n$$

$$S = n+\ldots+3+2+1$$

$$2S = n(n+1)$$

$$S = \frac{n(n+1)}{2}$$

$$\text{Sum} = \sum_{1}^{\infty} \frac{1}{2} \cdot \frac{n(n+1)}{n!}$$

$$= \frac{1}{2} \sum_{1}^{\infty} \left[ \frac{n}{(n-1)!} + \frac{1}{(n-1)!} \right]$$

$$= \frac{1}{2}[2e + e] \qquad \text{from previous proofs}$$

$$= \frac{3e}{2}$$

# Proof 37

$$\frac{1(2^2+1)}{2!}+\frac{2(3^2+1)}{3!}+\frac{3(4^2+1)}{4!}+\cdots=(3e+1)$$

$$\text{Sum}=\sum_{1}^{\infty}\left[\frac{n((n+1)^2+1)}{(n+1)!}\right]$$

$$=\sum_{1}^{\infty}\left[\frac{n^2(n+1)+n(n+1)+n}{(n+1)!}\right]$$

$$=\sum_{1}^{\infty}\left[\frac{n^2}{n!}+\frac{n}{n!}+\frac{n}{(n+1)!}\right]$$

$$=[2e+e+1]\qquad\text{from previous proofs}$$

$$=(3e+1)$$

For the sister series,

$$\frac{1(2^2-1)}{2!}+\frac{2(3^2-1)}{3!}+\frac{3(4^2-1)}{4!}+\cdots=3e-1$$

$$\text{Sum}=\sum_{1}^{\infty}\frac{(n-1)(n^2-1)}{n!}\qquad\text{(the first term}=0)$$

$$=\sum_{1}^{\infty}\frac{(n^3-n^2-n+1)}{n!}$$

$$=\sum_{1}^{\infty}\left[\frac{n^2}{(n-1)!}-\frac{n^2}{n!}-\frac{n}{n!}+\frac{1}{n!}\right]$$

$$=5e-2e-e+(e-1)\qquad\text{from previous proofs}$$

$$=(3e-1)$$

What beautiful symmetrical images for the sums of the two series. How amazing!

Can you prove the middle series (without the 1 in brackets) by yourself? It's simple.

# Proof 38

$$\frac{e^{\frac{1}{1}} \cdot e^{\frac{1}{3}} \cdot e^{\frac{1}{5}} \cdot e^{\frac{1}{7}} \cdot e^{\frac{1}{9}}}{e^{\frac{1}{2}} \cdot e^{\frac{1}{4}} \cdot e^{\frac{1}{6}} \cdot e^{\frac{1}{8}} \cdot e^{\frac{1}{10}}} \cdots = 2$$

From Proof 15, we saw that

$$\ln(1+x) = x - \frac{x^2}{2} + \frac{x^3}{3} - \frac{x^4}{4} + \cdots$$

Let $x = 1$,

$$\ln 2 = 1 - \frac{1}{2} + \frac{1}{3} - \frac{1}{4} + \cdots$$

Raising both sides of the equation as the powers of e,

$$e^{\ln 2} = e^{(1 - \frac{1}{2} + \frac{1}{3} - \frac{1}{4} \cdots)}$$

$$2 = \frac{e^{(1 + \frac{1}{3} + \frac{1}{5} + \cdots)}}{e^{(\frac{1}{2} + \frac{1}{4} + \frac{1}{6} \cdots)}}$$

$$= \frac{e^{\frac{1}{1}} \cdot e^{\frac{1}{3}} \cdot e^{\frac{1}{5}} \cdot e^{\frac{1}{7}} \cdot e^{\frac{1}{9}}}{e^{\frac{1}{2}} \cdot e^{\frac{1}{4}} \cdot e^{\frac{1}{6}} \cdot e^{\frac{1}{8}} \cdot e^{\frac{1}{10}}} \cdots$$

# Proof 39

$$e^{i\pi} = -1$$

$$e^{i\pi} + 1 = 0$$

The above equation — "The Most Beautiful Equation in the World" — is also called "Euler's Identity", and can be written in two forms. I prefer the former as it is shorter, and the (−1) on the right hand side alludes to the magic of the imaginary number $i$, where $i^2$ is equal to (−1). Euler started his proof (published in his classic "*Introductio in analysin infinitorum*" in 1748) with $e^x$:

$$e^x = 1 + \frac{x}{1!} + \frac{x^2}{2!} + \frac{x^3}{3!} + \frac{x^4}{4!} + \cdots$$

Substituting $x$ by ($ix$), assuming its validity, he obtained:

$$e^{ix} = 1 + \frac{ix}{1!} + \frac{(ix)^2}{2!} + \frac{(ix)^3}{3!} + \frac{(ix)^4}{4!} + \cdots$$

Remembering that $i^2 = -1$, $i^3 = -i$, and $i^4 = +1$;

$$e^{ix} = 1 + \frac{ix}{1!} - \frac{x^2}{2!} - \frac{ix^3}{3!} + \frac{x^4}{4!} + \cdots$$

$$= \left(1 - \frac{x^2}{2!} + \frac{x^4}{4!} - \frac{x^6}{6!} + \frac{x^8}{8!} - \cdots\right)$$

$$+ i\left(\frac{x}{1!} - \frac{x^3}{3!} + \frac{x^5}{5!} - \cdots\right)$$

From the "Elementary Series" in the Appendix, we know that the two series in brackets are the cosine and sine series, respectively.

$$\therefore \quad e^{ix} = \cos x + i \sin x$$

Let $x = \pi$:

$$e^{i\pi} = \cos \pi + i \sin \pi$$

$$= -1 + 0$$

$$= -1$$

When Euler first derived this equation in the 18th century, there was a lack of understanding of the concept of convergence. Hence he rearranged the terms of his $e^{ix}$ series liberally. Later developments in mathematics led to a better understanding of convergence, and to the concept of "circle of convergence" for complex numbers involving $i$. Advanced mathematics since then has confirmed the validity of Euler's creation. Hence he rightly deserves the honour of "Euler's Identity".

Euler's equation

$$e^{ix} = \cos x + i \sin x$$

gives rise to a number of very unexpected (surprising and incredible) results: e.g., For $x = \frac{\pi}{2}$

$$e^{ix} = e^{i\frac{\pi}{2}} = \cos\left(\frac{\pi}{2}\right) + i \sin\left(\frac{\pi}{2}\right)$$

$$= 0 + i$$

$$= i$$

Noting that "power" in mathematics means multiplying a number by itself for the appropriate number of times, viz:

$$1^1 = 1$$
$$2^2 = 2 \times 2$$
$$3^3 = 3 \times 3 \times 3$$
$$\cdots\cdots\cdots\cdots$$
$$\cdots\cdots\cdots\cdots$$

the question arises — what is "the meaning of $i^i$ ($i$ raised to the power of $i$)?

There is no "real" meaning to $i^i$.

Nevertheless, mathematicians have explored the concept.

Would $i^i$ give any result at all?

An imaginary part of a complex number for the answer?

Or a complex number?

Or even a real number (i.e. without $i$)?

Would it be stretching the imagination if $i^i$ turns out to be an infinite number of real numbers?

Prepare yourself for the surprise of your life — and turn the page.

Earlier on, we saw that:

$$i = e^{i\frac{\pi}{2}}$$

$$\therefore \quad i^{i} = (e^{i\frac{\pi}{2}})^{i}$$

$$= e^{i^{2} \cdot \frac{\pi}{2}}$$

$$= e^{-\frac{\pi}{2}}$$

$$= \frac{1}{e^{\frac{\pi}{2}}}$$

$$= 0.2079 \ldots$$

We further note that

$i = e^{i\frac{\pi}{2}}$ is just one of an infinite number of values for the equation.

$$e^{ix} = \cos x + i \sin x$$

since $\sin \dfrac{\pi}{2}$, $\sin \dfrac{5\pi}{2}$, $\sin \dfrac{9\pi}{2} \cdots$ are all equal to 1.

$\therefore$ The general equation is:

$$i = e^{i(4n+1)\frac{\pi}{2}}$$

For $\quad n = 0, \quad i^{i} = 0.2079 \ldots$

$\quad n = -1, \quad i^{i} = 111.3178 \ldots$

$\quad n = 1, \quad i^{i} = 3.882 \ldots \times 10^{-4}$

$\quad n = 2, \quad i^{i} = 7.249 \ldots \times 10^{-7}$

$\cdots\cdots\cdots\cdots\cdots\cdots\cdots\cdots\cdots\cdots$

$\cdots\cdots\cdots\cdots\cdots\cdots\cdots\cdots\cdots\cdots$

Yes, Leonhard Euler was the first mathematician to work out that $i^{i}$ is a real number, and that there are an infinite number of values for this most mysterious entity in mathematics!

# Not-So-Easy Proofs

I had a feeling once about Mathematics —
that I saw it all
Depth upon depth was revealed to me —
a quantity passing through infinity
and changing its sign from plus to minus
I saw exactly why it happened ...
but as it was after dinner
I let it go

**Winston Churchill** (1874–1965)

§

Don't worry about your difficulty in mathematics
I can assure you mine are greater

**Albert Einstein** (1879–1955)

§

# Proof 40

$$1 - \frac{1}{3} + \frac{1}{5} - \frac{1}{7} + \frac{1}{9} - \cdots = \frac{\pi}{4}$$

Let $\tan y = x$; then $y = \arctan x$

$$\frac{d}{dx}(\tan y) = 1$$

$$\sec^2 y \cdot \frac{dy}{dx} = 1 \quad \text{(from "Elementary calculus")}$$

$$\frac{dy}{dx} = \frac{1}{\sec^2 y}$$

$$= \frac{1}{1 + x^2} \quad \text{(from "Elementary calculus")}$$

From division,

$$\frac{1}{1 + x^2} = 1 - x^2 + x^4 - x^6 + \ldots$$

Integrating both sides of the equation:

$$\int_1^\infty \frac{dy}{dx} \cdot dx = \int_1^\infty \frac{1}{1 + x^2} \cdot dx = \int_1^\infty (1 - x^2 + x^4 - x^6 \ldots) \cdot dx$$

$$y = x - \frac{x^3}{3} + \frac{x^5}{5} - \frac{x^7}{7} + \cdots$$

$$\arctan x = x - \frac{x^3}{3} + \frac{x^5}{5} - \frac{x^7}{7} + \cdots$$

Let $x = 1$; then $\arctan(1) = \frac{\pi}{4}$

$$\therefore \quad \frac{\pi}{4} = 1 - \frac{1}{3} + \frac{1}{5} - \frac{1}{7} + \frac{1}{9} - \cdots$$

# Proof 41

$$\frac{1}{1^2} + \frac{1}{2^2} + \frac{1}{3^2} + \frac{1}{4^2} + \cdots = \frac{\pi^2}{6}$$

By simple binomial expansion,

$$(a+x)^2 = a^2 + 2ax + x^2$$
$$(a+x)^3 = a^3 + 3a^2x + 3ax^2 + x^3$$
$$\cdots\cdots\cdots\cdots\cdots\cdots\cdots\cdots\cdots\cdots$$
$$\cdots\cdots\cdots\cdots\cdots\cdots\cdots\cdots\cdots\cdots$$
$$(a+x)^n = a_0 + a_1x + a_2x^2 + \cdots + x^n$$

where $a_0$, $a_1$, $a_2$, $a_3$ ... are constants, the coefficients of the different terms in the expansion.

Similarly, a general function $f(x)$ with different roots can be written in the same way:

$$f(x) = a_0 + a_1x + a_2x^2 + \cdots + x^n$$

Writing this function in the form of products of its different factors, we get:

$$f(x) = (x - r_1)(x - r_2)(x - r_3)...(x - r_n)$$

where $r_1$, $r_2$, $r_3$ ... $r_n$ are the roots of the equation.

Dividing the equation by $r_1 r_2 r_3$ ... $r_n$, and rearranging the terms, we can write the equation in a new form:

$$f(x) = \left(1 - \frac{x}{r_1}\right)\left(1 - \frac{x}{r_2}\right)\left(1 - \frac{x}{r_3}\right)\cdots\left(1 - \frac{x}{r_n}\right)$$

From "Elementary Trigonometry" the function

$$\frac{\sin x}{x} = 1 - \frac{x^2}{3!} + \frac{x^4}{5!} - \frac{x^6}{7!} + \cdots$$

Following from the earlier equation, we can write this equation in the form of products of factors:

$$\frac{\sin x}{x} = \left(1 - \frac{x^2}{r_1}\right)\left(1 - \frac{x^2}{r_2}\right)\left(1 - \frac{x^2}{r_3}\right)\cdots$$

Hence,

$$1 - \frac{x^2}{3!} + \frac{x^4}{5!} - \frac{x^6}{7!} \cdots = \left(1 - \frac{x^2}{r_1}\right)\left(1 - \frac{x^2}{r_2}\right)\cdots$$

Multiplying out the RHS as an expansion and equating the coefficients of the corresponding terms, we get the coefficients of $x^2$:

$$-\frac{1}{3!} = -\left(\frac{1}{r_1} + \frac{1}{r_2} + \cdots\right)$$

For $\sin x = 0$, $x = \pi,\ 2\pi,\ 3\pi\ \ldots$

since $\sin\pi$, $\sin 2\pi$, $\sin 3\pi \ldots$ are all equal to zero.

For $\dfrac{\sin x}{x} = 0$, $\left(1 - \dfrac{\pi^2}{r_1}\right) = 0$; $\left(1 - \dfrac{(2\pi)^2}{r_2}\right) = 0 \ldots$

$\therefore\ r_1 = \pi^2;\ r_2 = (2\pi)^2;\ r_3 = (3\pi)^2 \ldots$

$$\therefore\ -\frac{1}{3!} = -\left(\frac{1}{\pi^2} + \frac{1}{(2\pi)^2} + \frac{1}{(3\pi)^2} + \cdots\right)$$

$$\frac{1}{6} = \frac{1}{\pi^2}\left(\frac{1}{1^2} + \frac{1}{2^2} + \frac{1}{3^2} + \cdots\right)$$

$$\frac{\pi^2}{6} = \frac{1}{1^2} + \frac{1}{2^2} + \frac{1}{3^2} + \cdots$$

# Proof 42

$$\frac{1}{1^2} + \frac{1}{3^2} + \frac{1}{5^2} + \cdots \qquad = \frac{\pi^2}{8}$$

$$\frac{1}{2^2} + \frac{1}{4^2} + \frac{1}{6^2} + \cdots \qquad = \frac{\pi^2}{24}$$

$$\frac{1}{1^2} - \frac{1}{2^2} + \frac{1}{3^2} - \frac{1}{4^2} + \cdots = \frac{\pi^2}{12}$$

Since the series for $\frac{\pi^2}{6}$ is "absolutely convergent", we can apply arithmetical operations to the individual terms of the series.

$$\text{Sum} = \frac{\pi^2}{6} = \frac{1}{1^2} + \frac{1}{2^2} + \frac{1}{3^2} + \frac{1}{4^2} + \cdots$$

$$= \left( \frac{1}{1^2} + \frac{1}{3^2} + \frac{1}{5^2} + \cdots \right) + \left( \frac{1}{2^2} + \frac{1}{4^2} + \frac{1}{6^2} + \cdots \right)$$

$$= \left( \frac{1}{1^2} + \frac{1}{3^2} + \frac{1}{5^2} + \cdots \right) + \frac{1}{4}\left( \frac{1}{1^2} + \frac{1}{2^2} + \frac{1}{3^2} + \cdots \right)$$

$$= \left( \frac{1}{1^2} + \frac{1}{3^2} + \frac{1}{5^2} + \cdots \right) + \frac{1}{4}\,\text{Sum}$$

$$\therefore \quad \frac{3}{4}\,\text{Sum} = \frac{3}{4} \cdot \frac{\pi^2}{6} = \frac{1}{1^2} + \frac{1}{3^2} + \frac{1}{5^2} + \cdots$$

$$\therefore \quad \frac{\pi^2}{8} = \frac{1}{1^2} + \frac{1}{3^2} + \frac{1}{5^2} + \cdots$$

$$\frac{1}{4}\frac{\pi^2}{6} = \frac{1}{4}\left( \frac{1}{1^2} + \frac{1}{2^2} + \frac{1}{3^2} + \cdots \right)$$

$$\therefore \quad \frac{\pi^2}{24} = \frac{1}{2^2} + \frac{1}{4^2} + \frac{1}{6^2} + \cdots$$

$$\frac{\pi^2}{8} - \frac{\pi^2}{24} = \left( \frac{1}{1^2} + \frac{1}{3^2} + \frac{1}{5^2} + \cdots \right) - \left( \frac{1}{2^2} + \frac{1}{4^2} + \frac{1}{6^2} + \cdots \right)$$

$$\therefore \quad \frac{\pi^2}{12} = \frac{1}{1^2} - \frac{1}{2^2} + \frac{1}{3^2} - \frac{1}{4^2} + \cdots$$

# Proof 43

$$\frac{\sqrt{2}}{2} \cdot \frac{\sqrt{2+\sqrt{2}}}{2} \cdot \frac{\sqrt{2+\sqrt{2+2}}}{2} \cdots = \frac{2}{\pi}$$

From "Elementary Trigonometry", we get:

$$\sin 2x = 2\cos x \sin x$$

$$\therefore \quad \sin x = 2\cos\frac{x}{2}\sin\frac{x}{2}$$

$$= 2\cos\frac{x}{2} \cdot 2\cos\frac{x}{2^2}\sin\frac{x}{2^2}$$

$$\cdots\cdots\cdots\cdots\cdots\cdots\cdots\cdots\cdots$$

$$= 2^n\cos\frac{x}{2}\cos\frac{x}{2^2}\cdots\cos\frac{x}{2^n}\cdot\sin\frac{x}{2^n}$$

Dividing both sides by $x$ and rearranging, we get:

$$\frac{\sin x}{x} = \cos\frac{x}{2}\cos\frac{x}{2^2}\cdots\cos\frac{x}{2^n}\cdot\frac{\sin\dfrac{x}{2^n}}{\dfrac{x}{2^n}}$$

When $n \to \infty$, $2^n \to \infty$

$$\frac{x}{2^n} \to 0$$

From "Elementary Trigonometry":

$$\frac{\sin x}{x} \to 1 \quad \text{when } x \to 0$$

$$\therefore \quad \frac{\sin\dfrac{x}{2^n}}{\dfrac{x}{2^n}} \to 1 \quad \text{when } n \to \infty$$

Therefore the infinite series for $\frac{\sin x}{x}$ can now be written as:

$$\frac{\sin x}{x} = \cos\frac{x}{2}\cos\frac{x}{2^2}\cos\frac{x}{2^3}\cdots$$

For $x = \frac{\pi}{2}$,

$$\sin x = \sin\frac{\pi}{2} = 1$$

$$\cos\frac{x}{2} = \cos\frac{\pi}{4} = \frac{\sqrt{2}}{2}$$

$$\cos\frac{x}{2^2} = \sqrt{\frac{1+\cos\frac{\pi}{4}}{2}} \quad \text{(from "Elementary Trigonometry")}$$

$$= \frac{\sqrt{2+\sqrt{2}}}{2}$$

$$\cdots\cdots\cdots\cdots\cdots\cdots\cdots\cdots\cdots$$

$$\therefore \quad \frac{\sin x}{x} = \frac{1}{\frac{\pi}{2}}$$

$$= \frac{2}{\pi} = \frac{\sqrt{2}}{2}\frac{\sqrt{2+\sqrt{2}}}{2}\frac{\sqrt{2+\sqrt{2+\sqrt{2}}}}{2}\cdots$$

# Proof 44

$$\frac{2 \cdot 2}{1 \cdot 3} \cdot \frac{4 \cdot 4}{4 \cdot 4} \cdot \frac{6 \cdot 6}{5 \cdot 7} \cdots = \frac{\pi}{2}$$

Using the equation derived earlier (Proof 41),

$$\frac{\sin x}{x} = \left(1 - \frac{x^2}{\pi^2}\right)\left(1 - \frac{x^2}{(2\pi)^2}\right)\left(1 - \frac{x^2}{(3\pi)^2}\right)\cdots$$

For $x = \dfrac{\pi}{2}$, $\dfrac{\sin x}{x} = \dfrac{\sin\left(\dfrac{\pi}{2}\right)}{\left(\dfrac{\pi}{2}\right)}$

$$= \frac{1}{\left(\dfrac{\pi}{2}\right)}$$

$$= \frac{2}{\pi}$$

For $x = \frac{\pi}{2}$, the individual terms on the RHS of the equation are:

$$\left(1 - \frac{x^2}{\pi^2}\right) = \left(1 - \frac{\left(\dfrac{\pi}{2}\right)^2}{\pi^2}\right) = \left(1 - \frac{1}{4}\right) = \left(\frac{1 \cdot 3}{2 \cdot 2}\right)$$

$$\left(1 - \frac{x^2}{(2\pi)^2}\right) = \left(1 - \frac{\left(\dfrac{\pi}{2}\right)^2}{(2\pi)^2}\right) = \left(1 - \frac{1}{16}\right) = \left(\frac{3 \cdot 5}{4 \cdot 4}\right)$$

$$\therefore \quad \frac{2}{\pi} = \frac{1 \cdot 3}{2 \cdot 2} \cdot \frac{3 \cdot 5}{4 \cdot 4} \cdot \frac{5 \cdot 7}{6 \cdot 6} \cdots$$

or

$$\frac{\pi}{2} = \frac{2 \cdot 2}{1 \cdot 3} \cdot \frac{4 \cdot 4}{3 \cdot 5} \cdot \frac{6 \cdot 6}{5 \cdot 7}$$

# Proof 45

$$\frac{3\cdot3}{2\cdot4}\cdot\frac{6\cdot6}{5\cdot7}\cdot\frac{9\cdot9}{8\cdot10}\cdots = \frac{2\pi}{3\sqrt{3}}$$

$$\frac{4\cdot4}{3\cdot5}\cdot\frac{8\cdot8}{7\cdot9}\cdot\frac{12\cdot12}{11\cdot13}\cdots = \frac{\pi}{2\sqrt{2}}$$

$$\frac{6\cdot6}{5\cdot7}\cdot\frac{12\cdot12}{11\cdot13}\cdot\frac{18\cdot18}{17\cdot19}\cdots = \frac{\pi}{3}$$

Using the same equation as in Proof 44,

For $x = \frac{\pi}{3}$,

$$\frac{\sin x}{x} = \frac{\frac{\sqrt{3}}{2}}{\frac{\pi}{3}} = \frac{3\sqrt{3}}{2\pi}$$

The terms on the RHS of the equation are now:

$$\left(1 - \frac{\left(\frac{\pi}{3}\right)^2}{\pi^2}\right) = \left(1 - \frac{1}{3^2}\right) = \frac{8}{9} = \frac{2\cdot4}{3\cdot3}$$

$$\left(1 - \frac{\left(\frac{\pi}{3}\right)^2}{(2\pi)^2}\right) = \left(1 - \frac{1}{6^2}\right) = \frac{35}{36} = \frac{5\cdot7}{6\cdot6}$$

............................................

............................................

$$\therefore \quad \frac{2\pi}{3\sqrt{3}} = \frac{3\cdot3}{2\cdot4}\cdot\frac{6\cdot6}{5\cdot7}\cdot\frac{9\cdot9}{8\cdot10}$$

Similarly for $x = \frac{\pi}{4}$,

$$\frac{\pi}{2\sqrt{2}} = \frac{4 \cdot 4}{3 \cdot 5} \cdot \frac{8 \cdot 8}{7 \cdot 9} \cdot \frac{12 \cdot 12}{11 \cdot 13} \cdots$$

and for $x = \frac{\pi}{6}$,

$$\frac{\pi}{3} = \frac{6 \cdot 6}{5 \cdot 7} \cdot \frac{12 \cdot 12}{11 \cdot 13} \cdot \frac{18 \cdot 18}{17 \cdot 19} \cdots$$

# Proof 46

$$1 - \frac{1}{3}\left(\frac{1}{3}\right) + \frac{1}{5}\left(\frac{1}{3^2}\right) - \frac{1}{7}\left(\frac{1}{3^3}\right) + \cdots = \frac{\pi}{2\sqrt{3}}$$

From Proof 40,

$$\arctan x = x - \frac{x^3}{3} + \frac{x^5}{5} - \frac{x^7}{7} + \cdots$$

For $x = \frac{1}{\sqrt{3}}$, $\tan \frac{\pi}{6} = \frac{1}{\sqrt{3}}$; $\arctan\left(\frac{1}{\sqrt{3}}\right) = \frac{\pi}{6}$

Substituting in the equation:

$$\frac{\pi}{6} = \frac{1}{\sqrt{3}} - \frac{1}{3} \cdot \frac{1}{(\sqrt{3})^3} + \frac{1}{5} \frac{1}{(\sqrt{3})^5} - \cdots$$

$$= \frac{1}{\sqrt{3}}\left(1 - \frac{1}{3} \cdot \frac{1}{3} + \frac{1}{5} \frac{1}{3^2} - \cdots\right)$$

$$\frac{\pi}{2\sqrt{3}} = 1 - \frac{1}{3 \cdot 3} + \frac{1}{5 \cdot 3^2} - \frac{1}{7 \cdot 3^3} + \cdots$$

or

$$\frac{\pi}{2\sqrt{3}} = 1 - \frac{1}{3}\left(\frac{1}{3}\right) + \frac{1}{5}\left(\frac{1}{3^2}\right) - \frac{1}{7}\left(\frac{1}{3^3}\right) + \cdots$$

# Proof 47

$$\arctan\left(\frac{1}{n}\right) = \arctan\frac{1}{(n+1)} + \arctan\left(\frac{1}{n(n+1)+1}\right)$$

$$\arctan\left(\frac{1}{1}\right) = \arctan\left(\frac{1}{2}\right) + \arctan\left(\frac{1}{3}\right)$$

From "Elementary Trigonometry",

$$\tan(a-b) = \frac{\tan a - \tan b}{1 + \tan a \, \tan b}$$

Let $\tan a = A$; $\tan b = B$

$\therefore$ $\arctan A = a$; $\arctan B = b$

$$\tan(a-b) = \frac{A-B}{1+AB}$$

$$(a-b) = \arctan\left(\frac{A-B}{1+AB}\right)$$

$$\arctan A - \arctan B = \arctan\left(\frac{A-B}{1+AB}\right)$$

Let $A = \frac{1}{n}$, $B = \frac{1}{n+1}$; then

$$\arctan\left(\frac{1}{n}\right) - \arctan\left(\frac{1}{n+1}\right) = \arctan\left(\frac{\frac{1}{n} - \frac{1}{n+1}}{1 + \frac{1}{n} \cdot \frac{1}{n+1}}\right)$$

$$= \arctan\left(\frac{1}{n(n+1)+1}\right)$$

$$\therefore \quad \arctan\left(\frac{1}{n}\right) = \arctan\left(\frac{1}{n+1}\right) + \arctan\left(\frac{1}{n(n+1)+1}\right)$$

For $n = 1$,

$$\arctan\left(\frac{1}{1}\right) = \arctan\left(\frac{1}{2}\right) + \arctan\left(\frac{1}{3}\right)$$

# Proof 48

$$4 \arctan\left(\frac{1}{5}\right) - \arctan\left(\frac{1}{239}\right) = \frac{\pi}{4}$$

From "Elementary Trigonometry",

$$\tan 2x = \frac{2\tan x}{1 - \tan^2 x}$$

For $\tan x = (\frac{1}{5})$, $\arctan(\frac{1}{5}) = x$

$$\tan 2x = \frac{\dfrac{2}{5}}{1 - \left(\dfrac{1}{5}\right)^2}$$

$$= \frac{5}{12}$$

$$\tan 4x = \frac{2\tan 2x}{1 - \tan^2 2x}$$

$$= \frac{\dfrac{10}{12}}{1 - \left(\dfrac{5}{12}\right)^2}$$

$$= \frac{120}{119}$$

$$\tan(A - B) = \frac{\tan A - \tan B}{1 + \tan A\ \tan B}$$

Let $A = 4x$ and $B = \frac{\pi}{4}$

then

$$\tan\left(4x - \frac{\pi}{4}\right) = \frac{\tan 4x - \tan\left(\frac{\pi}{4}\right)}{1 + \tan 4x \tan\left(\frac{\pi}{4}\right)}$$

$$= \frac{\frac{120}{119} - 1}{1 + \frac{120}{119}} = \frac{1}{239}$$

$$4x - \frac{\pi}{4} = \arctan\left(\frac{1}{239}\right)$$

$$\therefore \quad 4\arctan\left(\frac{1}{5}\right) - \arctan\left(\frac{1}{239}\right) = \frac{\pi}{4}$$

# Proof 49

$$1 + \frac{1}{3 \cdot 2^3} + \frac{3}{4} \cdot \frac{1}{5 \cdot 2^5} + \frac{3 \cdot 5}{4 \cdot 6} \cdot \frac{1}{7 \cdot 2^7} \cdots = \frac{\pi}{3}$$

From "Elementary Trigonometry" and "Elementary Calculus",

$$\sin^2 y + \cos^2 y = 1$$
$$\cos y = \sqrt{1 - \sin^2 y}$$

Let $\sin y = x$; then $y = \arcsin x$

$$\frac{d}{dx}(\sin y) = 1$$

$$\cos y \cdot \frac{dy}{dx} = 1$$

$$\frac{dy}{dx} = \frac{1}{\cos y}$$

$$= \frac{1}{\sqrt{1 - \sin^2 y}}$$

$$= \frac{1}{\sqrt{(1 - x^2)}}$$

By expansion:

$$\frac{1}{\sqrt{(1 - x^2)}} = 1 + \frac{x^2}{2} + \frac{3}{8}x^4 + \frac{5}{16} \cdot x^6 + \ldots$$

Integrating both sides,

$$\int_1^\infty \frac{dy}{dx}\cdot dx = \int_1^\infty \frac{1}{\sqrt{1-x^2}}\cdot dx = \int_1^\infty \left(1+\frac{x^2}{2}+\frac{3}{8}x^4+\frac{5}{16}x^6\right)\cdot dx$$

$$y = x + \frac{1}{2}\frac{x^3}{3}+\frac{3}{8}\cdot\frac{x^5}{5}+\frac{5}{16}\frac{x^7}{7}+\cdots$$

$$\arcsin x = x + \frac{1}{2}\frac{x^3}{3}+\frac{1\cdot 3}{2\cdot 4}\frac{x^5}{5}+\frac{1\cdot 3\cdot 5}{2\cdot 4\cdot 6}\frac{x^7}{7}+\cdots$$

Let $x = \frac{1}{2}$, $\arcsin(\frac{1}{2}) = \frac{\pi}{6}$

$$\therefore\quad \frac{\pi}{6} = \frac{1}{2}+\frac{1}{2}\cdot\frac{1}{3\cdot 2^3}+\frac{1\cdot 3}{2\cdot 4}\cdot\frac{1}{5\cdot 2^5}+\frac{1\cdot 3\cdot 5}{2\cdot 4\cdot 6}\cdot\frac{1}{7\cdot 2^7}+\cdots$$

and

$$\frac{\pi}{3} = 1+\frac{1}{3\cdot 2^3}+\frac{3}{4}\cdot\frac{1}{5\cdot 2^5}+\frac{3\cdot 5}{4\cdot 6}\cdot\frac{1}{7\cdot 2^7}+\cdots$$

# Proof 50

$$\frac{1}{1}\left(\frac{1}{2}+\frac{1}{3}\right)-\frac{1}{3}\left(\frac{1}{2^3}+\frac{1}{3^3}\right)+\frac{1}{5}\left(\frac{1}{2^5}+\frac{1}{3^5}\right)-\cdots=\frac{\pi}{4}$$

$$\arctan\left(\frac{1}{1}\right)=\arctan\left(\frac{1}{2}\right)+\arctan\left(\frac{1}{3}\right)$$

$$\arctan x = x-\frac{x^3}{3}+\frac{x^5}{5}-\frac{x^7}{7}+\cdots$$

$$\arctan\left(\frac{1}{1}\right)=\frac{\pi}{4}$$

$$\arctan\left(\frac{1}{2}\right)=\left(\frac{1}{2}\right)-\frac{1}{3}\left(\frac{1}{2}\right)^3+\frac{1}{5}\left(\frac{1}{2}\right)^5-\cdots$$

$$\arctan\left(\frac{1}{3}\right)=\left(\frac{1}{3}\right)-\frac{1}{3}\left(\frac{1}{3}\right)^3+\frac{1}{5}\left(\frac{1}{3}\right)^5-\cdots$$

$$\therefore\quad \frac{\pi}{4}=\frac{1}{1}\left(\frac{1}{2}+\frac{1}{3}\right)-\frac{1}{3}\left(\frac{1}{2^3}+\frac{1}{3^3}\right)+\frac{1}{5}\left(\frac{1}{2^5}+\frac{1}{3^5}\right)-\cdots$$

# Proof 51

$$\frac{1}{4}\tan\left(\frac{\pi}{4}\right)+\frac{1}{8}\tan\left(\frac{\pi}{8}\right)+\frac{1}{16}\tan\left(\frac{\pi}{16}\right)+\cdots \quad =\frac{1}{\pi}$$

$$\frac{1}{2^2}\tan\left(\frac{\pi}{2^2}\right)+\frac{1}{2^3}\tan\left(\frac{\pi}{2^3}\right)+\frac{1}{2^4}\tan\left(\frac{\pi}{2^4}\right)+\cdots=\frac{1}{\pi}$$

From "Elementary Trigonometry",

$$\tan A = \frac{2\tan\left(\dfrac{A}{2}\right)}{1-\tan^2\left(\dfrac{A}{2}\right)}$$

Inverting the equation

$$\cot A = \frac{1-\tan^2\left(\dfrac{A}{2}\right)}{2\tan\left(\dfrac{A}{2}\right)}$$

$$=\frac{1}{2}\cot\left(\frac{A}{2}\right)-\frac{1}{2}\tan\left(\frac{A}{2}\right)$$

$$=\frac{1}{2^2}\cot\left(\frac{A}{2^2}\right)-\frac{1}{2^2}\tan\left(\frac{A}{2^2}\right)-\frac{1}{2}\tan\left(\frac{A}{2}\right)$$

$$\cdots\cdots\cdots\cdots\cdots\cdots\cdots\cdots\cdots\cdots\cdots\cdots\cdots\cdots\cdots\cdots\cdots$$

$$\cdots\cdots\cdots\cdots\cdots\cdots\cdots\cdots\cdots\cdots\cdots\cdots\cdots\cdots\cdots\cdots\cdots$$

$$=\frac{1}{2^n}\cot\left(\frac{A}{2^n}\right)-\cdots\frac{1}{2^2}\tan\left(\frac{A}{2^2}\right)-\frac{1}{2}\tan\left(\frac{A}{2}\right)$$

$$=\frac{1}{2^n}\cot\left(\frac{A}{2^n}\right)-\left(\frac{1}{2}\tan\left(\frac{A}{2}\right)+\frac{1}{2^2}\tan\left(\frac{A}{2^2}\right)\cdots\right)$$

Rearranging with $n \to \infty$,

$$\frac{1}{A} - \cot A = \frac{1}{2}\tan\frac{A}{2} + \frac{1}{4}\tan\frac{A}{4} + \frac{1}{8}\tan\frac{A}{8} + \cdots \qquad \text{(see footnote)}$$

Let $A = \frac{\pi}{4}$; $\therefore \cot A = \cot\left(\frac{\pi}{4}\right) = \tan\left(\frac{\pi}{4}\right)$

$$\frac{4}{\pi} - \tan\left(\frac{\pi}{4}\right) = \frac{1}{2}\tan\left(\frac{\pi}{8}\right) + \frac{1}{4}\tan\left(\frac{\pi}{16}\right) + \cdots$$

$$\frac{4}{\pi} = \tan\left(\frac{\pi}{4}\right) + \frac{1}{2}\tan\left(\frac{\pi}{8}\right) + \frac{1}{4}\tan\left(\frac{\pi}{16}\right) + \cdots$$

$$\frac{1}{\pi} = \frac{1}{4}\tan\left(\frac{\pi}{4}\right) + \frac{1}{8}\tan\left(\frac{\pi}{8}\right) + \frac{1}{16}\tan\left(\frac{\pi}{16}\right) + \cdots$$

$$\frac{1}{\pi} = \frac{1}{2^2}\tan\left(\frac{\pi}{2^2}\right) + \frac{1}{2^3}\tan\left(\frac{\pi}{2^3}\right) + \frac{1}{2^4}\tan\left(\frac{\pi}{2^4}\right) + \cdots$$

---

$$\frac{1}{2^n}\cot\frac{A}{2^n} = \frac{1}{A} \cdot \frac{A}{2^n} \cdot \cot\frac{A}{2^n}$$

$$= \frac{1}{A} \cdot \frac{A}{2^n} \frac{\cos\left(\dfrac{A}{2^n}\right)}{\sin\left(\dfrac{A}{2^n}\right)}$$

$$= \frac{1}{A}$$

because as

$$n \to \infty$$

$$\cos\frac{A}{2^n} \to \cos 0 = 1$$

and

$$\frac{\sin\left(\dfrac{A}{2^n}\right)}{\left(\dfrac{A}{2^n}\right)} \to 1$$

# *Appendix*

The Appendix contains elementary formulas used in this book without derivation. These formulas can be found in most standard textbooks in mathematics.

Whoever despises the high wisdom of mathematics
nourishes himself on delusions

**Leonardo da Vinci** (1452–1519)

§

To explain all nature is too difficult for any one man
or even for any one age
Tis better to do a little with certainty
and leave the rest for others that come after you

**Isaac Newton** (1642–1727)

§

# Elementary Trigonometry

$$\sin 0 \quad = \sin \pi \;= 0 \;= \frac{\sqrt{0}}{2} \;= \sqrt{\frac{0}{4}}$$

$$\sin 30° = \sin \frac{\pi}{6} \;= \frac{1}{2} = \frac{\sqrt{1}}{2} \;= \sqrt{\frac{1}{4}}$$

$$\sin 45° = \sin \frac{\pi}{4} \;= \quad= \frac{\sqrt{2}}{2} \;= \sqrt{\frac{2}{4}}$$

$$\sin 60° = \sin \frac{\pi}{3} \;= \quad= \frac{\sqrt{3}}{2} \;= \sqrt{\frac{3}{4}}$$

$$\sin 90° = \sin \frac{\pi}{2} \;= 1 \;= \frac{\sqrt{4}}{2} \;= \sqrt{\frac{4}{4}}$$

$$\cos 0 \quad = \cos \pi \;= 1 \;= \frac{\sqrt{4}}{2} \;= \sqrt{\frac{4}{4}}$$

$$\cos 30° = \cos \frac{\pi}{6} \;= \quad= \frac{\sqrt{3}}{2} \;= \sqrt{\frac{3}{4}}$$

$$\cos 45° = \cos \frac{\pi}{4} \;= \quad= \frac{\sqrt{2}}{2} \;= \sqrt{\frac{2}{4}}$$

$$\cos 60° = \cos \frac{\pi}{3} \;= \frac{1}{2} = \frac{\sqrt{1}}{2} \;= \sqrt{\frac{1}{4}}$$

$$\cos 90° = \cos \frac{\pi}{2} \;= 0 \;= \frac{\sqrt{0}}{2} \;= \sqrt{\frac{0}{4}}$$

$$\tan 0 \quad = \tan \pi \ = 0$$

$$\tan 30^\circ = \tan\frac{\pi}{6} = \frac{1}{\sqrt{3}}$$

$$\tan 45^\circ = \tan\frac{\pi}{4} = 1$$

$$\tan 60^\circ = \tan\frac{\pi}{3} = \frac{\sqrt{3}}{1}$$

$$\tan 90^\circ = \tan\frac{\pi}{2} \to \infty$$

$$\sin (A \pm B) = \sin A \, \cos B \pm \cos A \, \sin B$$

$$\sin 2A \quad = 2 \sin A \, \cos A$$

$$\sin A \quad = 2 \sin\frac{A}{2} \, \cos\frac{A}{2}$$

$$\cos (A \pm B) = \cos A \, \cos B \mp \sin A \, \sin B$$

$$\cos 2A \quad = \cos^2 A - \sin^2 A$$

$$= 2 \cos^2 A - 1$$

$$= 1 - 2 \sin^2 A$$

$$\cos A \quad = 2 \cos^2 \frac{A}{2} - 1$$

$$\cos^2 A + \sin^2 A = 1$$

$$\cos^2 A \qquad = 1 - \sin^2 A$$

$$\sin^2 A \qquad = 1 - \cos^2 A$$

$$\tan(A + B) = \frac{\tan A + \tan B}{1 - \tan A \, \tan B}$$

$$\tan 2A \quad = \frac{2 \tan A}{1 - \tan^2 A}$$

$$\tan(A - B) = \frac{\tan A - \tan B}{1 + \tan A \, \tan B}$$

# Elementary Series

$$\frac{1}{(1+x)} = 1 - x + x^2 - x^3 + x^4 - \cdots$$

$$\frac{1}{(1-x)} = 1 + x + x^2 + x^3 + x^4 + \cdots$$

$$\frac{1}{(1-x)^2} = 1 + 2x + 3x^2 + 4x^3 + 5x^4 + \cdots$$

$$\ln(1+x) = x - \frac{x^2}{2} + \frac{x^3}{3} - \frac{x^4}{4} + \cdots$$

$$\ln(1-x) = -\left( x + \frac{x^2}{2} + \frac{x^3}{3} + \frac{x^4}{4} + \cdots \right)$$

$$\sin x = \frac{x}{1!} - \frac{x^3}{3!} + \frac{x^5}{5!} - \frac{x^7}{7!} + \cdots$$

$$\frac{\sin x}{x} = \frac{1}{1!} - \frac{x^2}{3!} + \frac{x^4}{5!} - \frac{x^6}{7!} + \cdots$$

$$\frac{\sin x}{x} = \cos\frac{x}{2} \cos\frac{x}{4} \cos\frac{x}{8} \cdots$$

$$\cos x = 1 - \frac{x^2}{2!} + \frac{x^4}{4!} - \frac{x^6}{6!} + \cdots$$

$$\frac{\sin x}{x} = \left( 1 - \frac{x^2}{\pi^2} \right)\left( 1 - \frac{x^2}{(2\pi)^2} \right)\left( 1 - \frac{x^2}{(3\pi)^2} \right)\cdots$$

$$\cos x = \left( 1 - \frac{(2x)^2}{\pi^2} \right)\left( 1 - \frac{(2x)^2}{(3\pi)^2} \right)\left( 1 - \frac{(2x)^2}{(5\pi)^2} \right)\cdots$$

$$\text{arc sin } x = x - \frac{1}{2}\cdot\frac{x^3}{3} + \frac{1\cdot3}{2\cdot4}\cdot\frac{x^5}{5} - \frac{1\cdot3\cdot5}{2\cdot4\cdot6}\cdot\frac{x^7}{7} + \cdots$$

$$\text{arc tan } x = x - \frac{x^3}{3} + \frac{x^5}{5} - \frac{x^7}{7} + \cdots$$

# Elementary Calculus

$$\frac{d}{dx}(x) = 1$$

$$\frac{d}{dx}(x^2) = 2x$$

$$\frac{d}{dx}(e^x) = e^x \qquad\qquad \int e^x \cdot dx = e^x + c$$

$$\frac{d}{dx}(\ln x) = \frac{1}{x} \qquad\qquad \int \frac{1}{x} \cdot dx = \ln x + c$$

$$\frac{d}{dx}(\ln(1+x)) = \frac{1}{(1+x)} \qquad\qquad \int \frac{1}{(1+x)} \cdot dx = \ln(1+x) + c$$

$$\frac{d}{dx}(\sin x) = \cos x$$

$$\frac{d}{dx}(\cos x) = -\sin x$$

$$\frac{d}{dx}(\tan x) = \sec^2 x = 1 + \tan^2 x$$

I
was once a bottle of ink
Inky Dinky
Thinky Inky
(Infi-Nity
in a) Bottle of Ink

(With Apologies to)
**Edward Lear** (1818–1888)

§


Old Mathematicians never die
They just "tend to infinity"

**Anonymous**

§

# About the Author

Dr Y E O Adrian graduated from the University of Singapore with first class Honours in Chemistry in 1966, and followed up with a Master of Science degree in 1968.

He received his Master of Arts and his Doctor of Philosophy degrees from Cambridge University in 1970, and did post-doctoral research at Stanford University, California.

For his research, he was elected Fellow of Christ's College, Cambridge and appointed Research Associate at Stanford University in 1970.

His career spans fundamental and applied research and development, academia, and top appointments in politics and industry. His public service includes philanthropy and sports administration. Among his numerous awards are the Charles Darwin Memorial Prize, the Republic of Singapore's Distinguished Service Order, the International Olympic Committee Centenary Medal, and the Honorary Fellowship of Christ's College, Cambridge University.